AVIATION SAFETY

A Capital Disaster

Aviation Safety
A Capital Disaster

BEN LOCHARY

Copyright

 The author can be contacted at this email address: ben.airsafety@gmail.com.

Print ISBN: 979-8-234-07459-1

Full cover creation: Kris Miller
Modified charts: Ramesh Yarlagadda
Project manager: Diane Timmons via Upwork

Printed in the United States of America

Dedication

I want to dedicate this book to the families and loved ones of those who perished in this horrific accident, the First Responders who braved unfathomable conditions in hopes of finding survivors, and the many professionals at the National Transportation Safety Board for their uncompromising commitment to safety.

Table of Contents

Table of Figures

Introduction

ON JANUARY 29, 2025, AN ARMY HELICOPTER collided with an American Airlines regional jet that was preparing to land at the Ronald Reagan Washington National airport. Following the accident, officials were quick to assure the public that flying is safe. After all, there had not been a fatal airline accident in the United States since 2009, but that provides no comfort to the families and loved ones of the sixty-seven innocent souls who perished in this horrific accident.

Although the number of accidents has decreased in recent years, the number of incidents and close calls has risen. During an FAA Safety Summit in 2023, National Transportation Safety Board Chair Jennifer Homendy warned, "the absence of a fatality or accident doesn't mean the presence of safety. There is always more we can do."

This accident caught many people by surprise, but not some industry insiders. For years, air traffic controllers at the Ronald Reagan Washington National Airport raised safety concerns about close calls and traffic congestion. The warning signs were there, but no one would listen. The post-accident investigation revealed alarming statistics about close calls between passenger planes and Army helicopters near the airport. In fact, two of these occurred the day before this accident.

An air traffic controller, who was working in the control tower hours before the collision, told ABC News that nearly half of the new controllers that come to Washington National Airport to complete their training withdraw because they feel that conditions are unsafe. At times, arriving and departing flights are so closely spaced that as one airplane lifts off, the next airplane touches down on the same runway. They call this a "squeeze play."

Was this accident an anomaly or a sign that flying is not as safe as officials would like us to believe? As a retired airline pilot with thousands of flying hours including hundreds of landings at Washington National Airport, I want to answer that question, and explain the complexities of this accident, in terms that a non-pilot can understand. Simply put, the system that enabled this accident is broken and needs to be fixed.

Below is a link to an ABC news 60 Minutes interview with an Air Traffic Controller who was on duty in the DCA control tower on the day of the accident, entitled "Inside The Tower."

---

https://youtu.be/gJImdimSsFY

Chapter 1
Background

JANUARY 29, 2025, IS A DATE that I will always remember. As the sun set on that fateful evening, my wife and I settled in to watch TV while our puppy, Charlie, slept beside us on his cozy bed. Suddenly our show was interrupted by BREAKING NEWS! I thought to myself, oh no, what now. It had only been a few weeks since we learned of the catastrophic wildfires in California, a few months since hurricane Helene devastated parts of western North Carolina. Then, I heard three words that riveted my attention, Washington National Airport. My heart sank as reports of a midair collision surfaced. My worst fears were realized when authorities confirmed that a helicopter collided with a passenger jet that was attempting to land on runway 33, a runway that I have landed on many times.

Sixty-seven lives were lost in this horrific and avoidable accident which took place in clear view of the air traffic controllers who were tasked with preventing such a tragedy. How is that possible? This accident was the result of human error which was enabled by a culture that prioritizes profit over safety, a government bureaucracy that is paralyzed by its own regulations, and a Congress that is too polarized to enact reasonable legislation. I want to explain the complexities of this accident in terms that a non-pilot can understand but first, I want to explain my connection to this airport.

I can remember, when I was very young, traveling to Washington National Airport with my family to drop off or pick up a passenger. Things were very different in those days. Anyone could enter the terminal, no ticket required and no security screening. Once inside the main terminal, my eyes were drawn to a wall of windows that offered a panoramic view of the airfield. Better yet, just beyond those windows was an observation deck for sightseers to enjoy. I could spend the entire day taking in the sights and sounds while leaning against the railing that separated the onlookers from the action. I still recall the rumble of those powerful radial engines as the behemoth propellor driven airplanes taxied past. I waved enthusiastically and the pilots waved back. As a bonus, the Bolling Air Force Base was located just across the Potomac River to the east and occasionally I would see a military plane taking off or landing. On the north side of the river stands the Washington Monument which seemed to tower above the departing airplanes as they began their gravity-defying ascent. Little did I know back then that one day I would be in command of one of those flying machines.

I made my first flight as a newly qualified airline pilot during the summer of 1978. As the sun was rising, I climbed into the copilot's seat of a Beechcraft Model 99 alongside Captain Jesse McCormick. Jesse was somewhat of a legend because he had more experience flying the 99 than most people have driving a car, so the new pilots flew with him first. We departed Salisbury, Maryland right on schedule for the brief thirty-minute flight to Washington National Airport. As we approached runway 33, we passed directly over the since abandoned Bolling Air Force Base. I could still make out some of the runways and taxiways, but urban development was quickly erasing that chapter of aviation history. Jesse guided that airplane to a perfect landing, as he always did, and that was my first of many flights to this storied airport. During the next thirty-eight years I logged hundreds of landings at the Washington National Airport.

My memories of the airport and the surrounding airspace are so vivid that I feel like this accident occurred in my backyard. The collision involved a regional jet that was preparing to land on runway 33 and a US Army helicopter that was conducting a training flight near the airport. The regional jet was operated by PSA Airlines which is a subsidiary of American Airlines. Its flight number is PSA 5342, but some reports refer to it as American 5342. The airplane that was involved in the accident was originally designed and built by the Canadian airplane manufacturer Canadair and is known as a Canadair Regional Jet or CRJ. Throughout this book I will refer to the airplane as PSA 5342 or the CRJ.

PSA 5342 was flown by Captain Jonathan Campos, and First Officer Samuel Lilley. The cabin crew for this flight consisted of Flight Attendants Ian Epstein and Danasia Elder. Sixty passengers from all walks of life were seated in the cabin, including business travelers, vacationers, seven coworkers returning from a hunting trip, and members of the figure skating community, their coaches, and some family members. Among those were 1994 World Figure Skating Champions Vadim Naumov and Evgenia Shishkova, the parents of Maxim Naumov. Twenty-three-year-old Max had returned home on an earlier flight. In a later interview Max said, "my mom always texts me and calls me as soon as they land." That call never came. One year after the accident, Max represented the United States in the 2026 Winter Olympic Games in Milan, Italy. Holding back tears as he waited for his scores, Max held a picture of his parents that he carried with him throughout the games. Sadly, this is not the first time that tragedy has struck the figure skating community. In 1961, all eighteen members of the US figure skating team, along with sixteen coaches, officials, and family members perished in a plane crash in Belgium while en route to the World Figure Skating Championships in Prague.

The helicopter that was involved in this collision was a Sikorsky UH-60L Blackhawk operated by the US Army Bravo Company, 12th Combat Aviation Battalion based at Davidson Army Airfield in Fort Belvoir, Virginia. The primary mission of this Army unit is to transport senior military officials and the call sign for the helicopter that was involved in this accident was PAT 25 (Priority Air Transport). Throughout this book I will refer to the helicopter involved in this accident as PAT 25 or the helicopter.

PAT 25 was flown by Captain Rebecca Lobach, and she was receiving pilot proficiency and night vision goggle proficiency training. The other pilot was Chief Warrant Officer Andrew Eaves and he was acting as the instructor pilot. Also onboard was the crew chief, Staff Sargeant Ryan O'Hara.

Chapter 2
The Airport

THE WASHINGTON NATIONAL AIRPORT was renamed the Ronald Reagan Washington National Airport in 1998 as a tribute to President Ronald Reagan. The airport code for this airport is DCA. The international airport code is KDCA. The "K" designates the contiguous United States. DCA is unique due to its location, size, and traffic congestion. Situated on 860 acres, 133 of which are submerged in the Potomac River, it is surrounded by sensitive areas such as the White House, the United States Capitol, and the Pentagon. Its outdated runway layout, traffic congestion, and airspace restrictions make it less than ideal for its current use. It is designated by the FAA as a "special airport" due to its complex arrival and departure procedures and pilots must meet special requirements to fly to this airport.

Legislation to build the airport was first proposed in 1927 but Congress could not agree on the location. Frustrated by the lack of progress, President Roosevelt intervened and personally chose the present-day site. Construction began in 1938 and nearly twenty million cubic yards of sand and gravel were hauled in to fill the marshland to support the runways. The airport opened in1941 shortly before the US was drawn in to World War II.

The original terminal building is still in use today. It is a stately-looking building in keeping with the magnificent architecture that defines our nation's capital, and it looks just as I remember from years past. The wall of windows still provides a panoramic view of the airfield, but the outdoor observation area is no longer accessible due to security concerns. The remaining buildings have been replaced in recent years. The runway configuration is largely unchanged except for the elimination of an east-west runway in 1956 and the addition of an airplane arresting system on both ends of the main runway in 2012. This arresting system, which is known as an Engineered Materials Arresting System or EMAS, consists of a bed of crushable material that is designed to stop an airplane that overshoots the end of a runway, similar to the way that a runaway truck ramp on a mountain highway safely stops an 18-wheeler that has lost its brakes. These systems are used when airports lack adequate space for traditional safety areas.

The runways at DCA were designed for propellor-driven airplanes which take off and land at relatively slow speeds, which is a good thing, but it comes at a cost. Airplanes that can take off and land at slow speeds typically cruise at slow speeds. The "jet age" has conditioned us to traveling long distances in short periods of time making prop planes obsolete. The aerodynamic characteristics that allow jet powered airplanes to fly fast also dictate that a higher speed is required for takeoff and landing. Longer runways are necessary to accommodate this increased speed. With rare exception, the north/south runway at Washington National is the only runway that is suitable for takeoff and landing by a passenger jet, and it leaves little room for error.

Runways are numbered from 1 through 36 for ease of identification. These numbers correspond to the runway's magnetic direction, rounded to the nearest ten degrees and not including the last zero. For example, runway 1 approximates a magnetic direction of 10 degrees and runway 33 approximates 330 degrees. Under normal conditions, runway 1, and its reciprocal, runway 19 are the only runways at DCA that are suitable for takeoff and landing by a passenger jet. The earth's magnetic field is slowly changing. Occasionally, this necessitates a change in runway identification. Prior to 1999, runway 1 at DCA aligned with a magnetic direction that was closer to 360 degrees, and it was identified as runway 36.

Pilots take off and land into the wind as much as possible so, the wind direction determines which runway is in use. An airplane flying at an air speed of 100 knots (100 nautical miles per hour) into a 10-knot headwind crosses the ground at 90 knots. Flying in the opposite direction with a 10-knot tailwind results in a ground speed of 110 knots, or 20 knots faster. Tailwinds drastically increase takeoff and landing distance, so they are rarely acceptable. When the wind blows across a runway it is called a crosswind, and pilots must use a special technique to compensate. You might think that an airplane that weighs seventy-five tons would not be affected by the wind but think of it this way. The wind is a mass of air that is moving across the earth's surface, and the airplane is suspended by that air mass and moves with it.

The wind at the time of this accident was blowing from the north-west at 15 to 25 knots resulting in a significant crosswind for the active runway, runway 1. Captain Campos agreed to land his CRJ on runway 33 to assist Air Traffic Control (ATC) with traffic congestion. Runway 33 is six football fields shorter than runway 1 but that is somewhat offset by landing with a headwind. However, landing on runway 33 presents additional challenges, which are central to this accident, as you will soon learn.

As you can see from the airport diagram in Figure 1, the runways and taxiways at DCA intersect at multiple locations. Take notice of the three circles labeled HS 1, HS 2, and HS 3. These are called "hot spots." A hot spot is a location on an airport with a history of, or potential for, collision or runway incursion, and where heightened attention by pilots is necessary. Hot spots remain in effect until such time that the risk has been reduced or eliminated but the hot spots at DCA appear to be permanent. Since the runways and taxiways are crammed into such a small area there is no way to eliminate these hazards. Recent examples of "near runway incursions" at DCA include an April 2024 incident when a Southwest flight was cleared to cross a runway while a Jet Blue flight was taking off from that runway. A controller was able to stop both airplanes in time to avert an accident. Just six weeks later an American Airlines flight was forced to abort its takeoff due to another airplane that was landing on a crossing runway. As you can see from Figure 2, there are no hot spots at the Washington Dulles International Airport (IAD) which was built on 12,000 acres of land, to address the pitfalls that plague DCA.

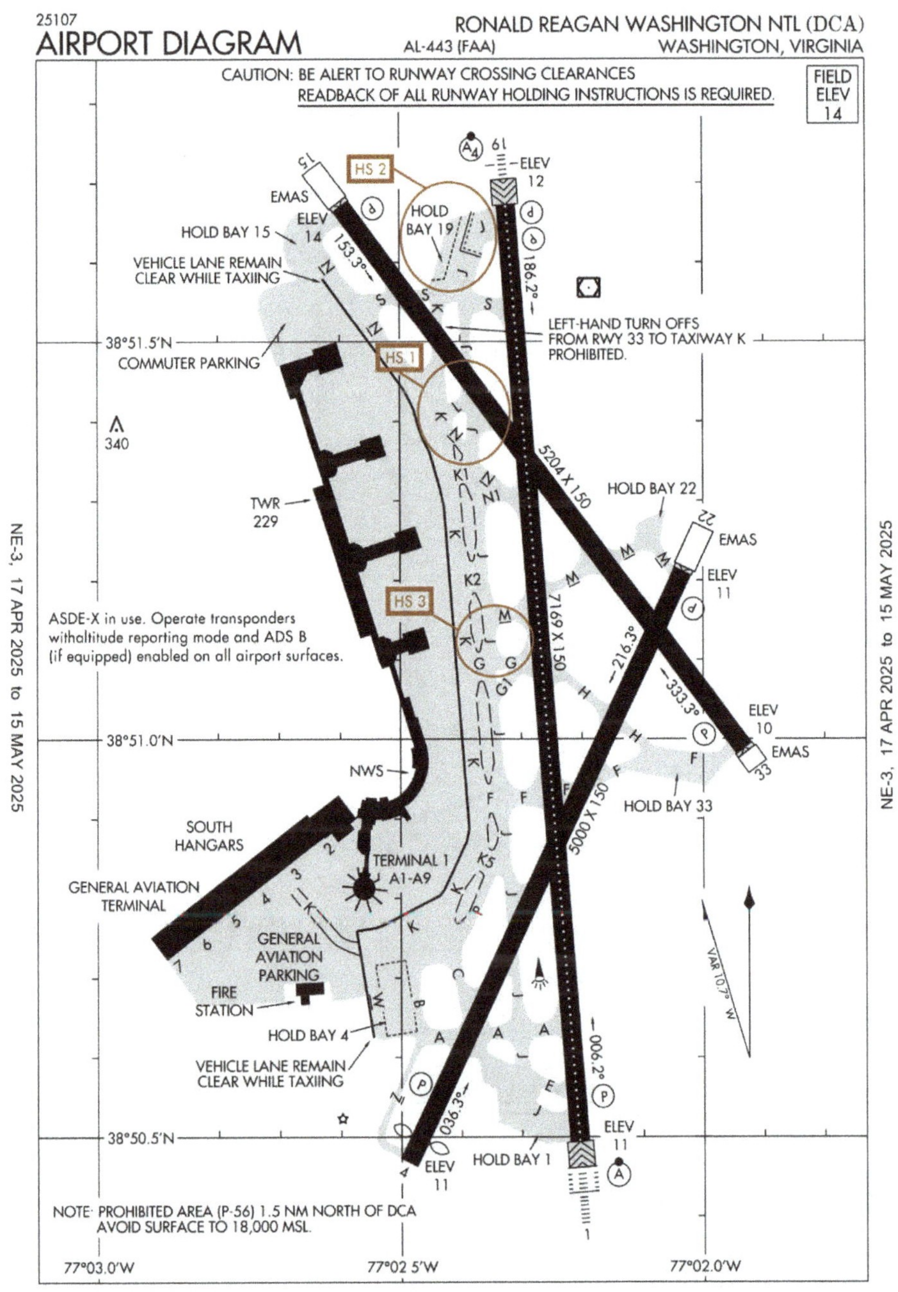

Figure 1. DCA Airport Diagram.

Source: FAA.gov

Note the complex runway intersections and hot spots.

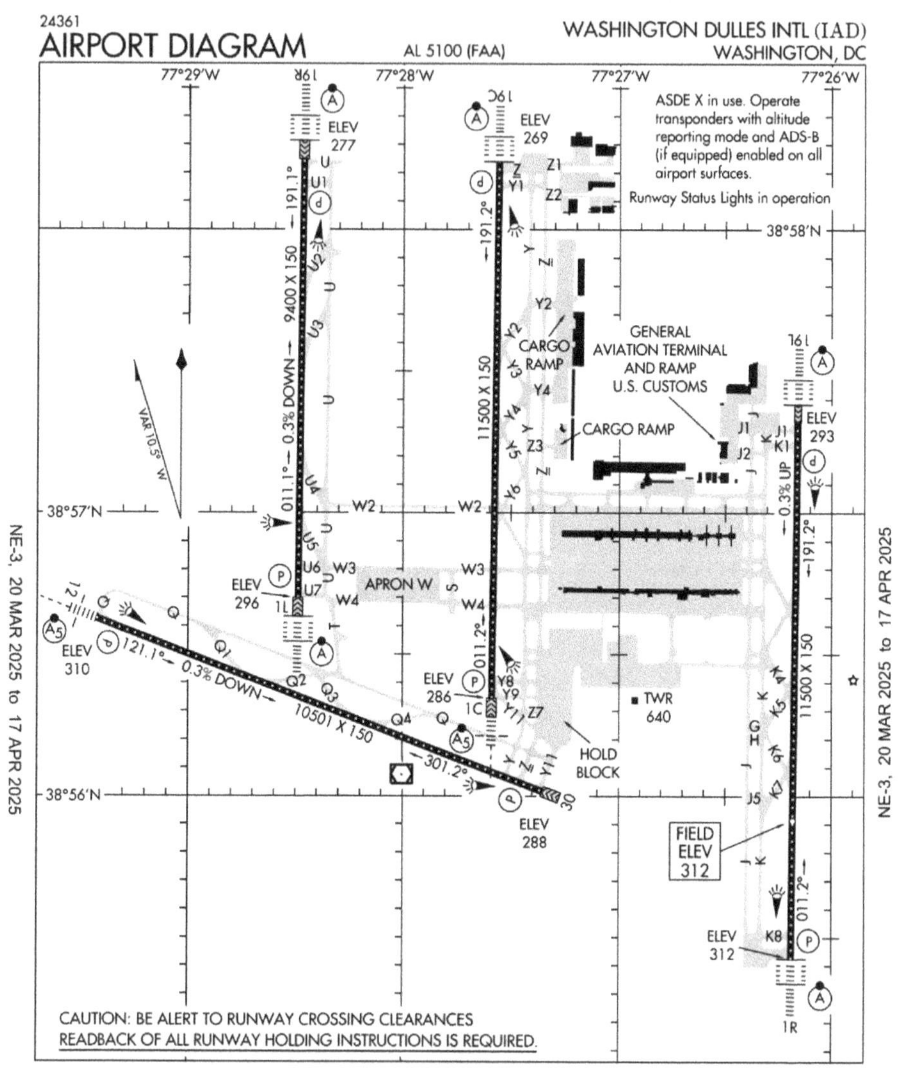

Figure 2. IAD Airport Diagram.

Source: FAA.gov

These runways are well spaced and accommodate simultaneous arrivals and departures with no need for hot spots.

To fit these diagrams on a standard size chart, they are drawn in different scales. The longest runway at IAD is sixty-three percent longer than the longest runway at DCA.

Even if more land was available to improve the runway layout at DCA, the airspace surrounding the airport presents additional challenges. When departing DCA to the north, an immediate turn is required after liftoff to avoid prohibited airspace followed by additional turns due to obstacle clearance and noise abatement procedures. Landing to the south requires the reverse of this convoluted path. By comparison, the nearby Washington Dulles International Airport (IAD) allows pilots to safely arrive and depart on a straight path to separate runways without the need for special procedures or hot spots.

The runway layout at DCA is inadequate for its current use but the airports' convenient location fuels its popularity. Proponents who continually demand more from DCA argue that the airport is safe because of the special procedures that apply but those procedures did not, and could not, prevent this accident. Runway limitations, airspace restrictions and traffic congestion test safety limits at this airport every day. Airport operations that far exceed airport capability combined with the unsafe practice of allowing helicopters to fly dangerously close to passenger planes, led to this accident.

To control airport congestion and reduce noise pollution, a "perimeter rule" was established in 1966. This rule limited flights to a 650-mile radius around DCA. A small number of existing flights that exceeded the 650-mile limit were allowed to continue. In later years, advocates for growth successfully lobbied to extend the radius to 1,250 miles. More recently, exceptions have been granted and there are currently forty flights, twenty round trips that exceed the 1,250-mile limit, some as far away as the west coast. Longer flights carry more fuel which increases takeoff weight pushing performance to the limit. Meanwhile, long-range flights have unlimited access to the nearby Washington Dulles International Airport (IAD), which is better suited to accommodate these flights.

To further control congestion DCA is one of five US airports that are subject to slot allocations. These slots limit the number of hourly takeoffs and landings, but once again, advocates for growth have successfully lobbied to increase the slot allocations on more than one occasion. The Washington National and Washington Dulles airports are federally owned and run by the Metropolitan Washington Airport Authority (MWAA) through a special lease agreement. According to MWAA, 90 percent of DCA's flights use its main runway making it the busiest runway in America with over 800 daily takeoffs and landings, which is a takeoff or landing every minute during most of the day. Pilots routinely rate DCA among the nation's most challenging airports. There is no room to expand. Its land is only a fraction of the area's other two commercial airports, Washington Dulles and Baltimore Washington International (BWI). It's just too small to fit all the flights that airlines want to send to Washington.

Nonetheless, special interest groups like the Capital Access Alliance, a coalition of business organizations including Delta Airlines, which incidentally is headquartered in Georgia and operates a connecting hub in Utah, continue to lobby for more and longer flights. In 2023 Congress men Hank Johnson (Georgia) and Burgess Owens

(Utah) introduced the Direct Capital Access Act, known as the DCA Act, to add 28 additional flights to DCA, some of which exceed the perimeter rule. Representative Owens stated, "when a federal regulation [the perimeter and slot rule] becomes so outdated that it unnecessarily hurts American consumers and arbitrarily burdens our nation's economic growth, it's time for Congress to act…Modernizing the perimeter rule will improve access to Washington, D.C., reduce airline ticket prices, and increase tax revenue for the area."

Airport overcrowding lies at the heart of this accident. The two Washington airports already provide nonstop flights to domestic and international destinations from Albany to Zurich. The huge Southwest Airlines hub at nearby BWI provides even more travel options. Furthermore, launching long range flights from the short runway at DCA pushes aircraft performance to the extreme as you will read in the next chapter. Although the Direct Capital Access Act did not pass, the relentless push for more flights continued, despite local opposition.

In a letter to the leaders of the House Committee on Transportation and Infrastructure, representatives from Virginia, the District of Columbia, and Maryland expressed strong opposition to changes in the perimeter and slot allocation rules at DCA. Highlights of this letter follow:

- DCA and IAD operate as an integrated system of federally owned assets and DCA was never intended to be a long-haul airport.
- DCA is designed to accommodate 15 million passengers annually but handled a record setting 24 million passengers last year (2023).

- DCA already experiences an above average number of missed approaches and additional flights would likely increase those numbers.

The full text of this letter is appended.

Just eight months prior to this tragic midair collision, Congress passed the FAA Reauthorization Act of 2024. Among other things, this bill aims to improve aviation safety, strengthen consumer protections, and modernize the national airspace system. Final passage of the bill hinged on one obstacle, increased takeoff and landing slots at DCA. The four senators from Maryland and Virginia opposed the bill arguing that additional flights went against safety concerns after a near-miss during the previous month and that it would increase delays at the already congested airport. Regardless, the bill passed, and 10 additional daily flights were added to the schedule.

According to a March 13, 2025, Associated Press report: "When Congress pushed ahead last year with adding ten new daily flights to Washington DC's Reagan National Airport, five of which exceed the perimeter rule, the Federal Aviation Administration had data showing an unnerving number of near misses in the already-crowded skies – something lawmakers apparently did not know."

In 1947, the CAA, predecessor to the FAA, predicted that DCA would exceed its capacity for both passengers and airplanes by 1955, seventy years prior to this accident.

Chapter 3
Aircraft Performance

FOR MANY PASSENGERS, THE DIFFERENCE between the Ronald Reagan Washington National Airport (DCA) and the Washington Dulles International Airport (IAD) is location, location, location. Travelers who reside in the DC metropolitan area have quick and easy access to DCA, but travel to IAD is less convenient. For me, the difference between these two airports is performance, performance, performance. To better understand, a brief explanation of aircraft performance is in order.

It is generally known that air becomes thinner, or less dense, as altitude increases. Anyone who has driven a car on mountain roads has experienced the sluggish response as their normally powerful engine gasps for oxygen at altitudes above sea level. Thin air has an even more pronounced effect on airplane performance. Not only do the airplane's engines produce less power as altitude increases, but the airplane's wings produce less lift. An airplane taking off from a sea-level airport will accelerate to flying speed and climb much faster than an identical airplane taking off in the thin air of a high-altitude airport. High-altitude airports, such as the Denver International Airport (DIA), have extremely long runways to compensate for the additional time that it takes for an airplane to accelerate to flying speed. The longest runway at DIA is more than twice the length of the longest runway at DCA, giving it the capability to launch heavily loaded long-range flights from an airport elevation of 5,500 feet above sea level.

You might wonder, what does this have to do with DCA, which is located at sea level. Air density is not only affected by altitude but also by temperature. As temperature increases, air density decreases. An airplane taking off on a cold winter day will accelerate to flying speed and climb much faster than an identical airplane departing from that same runway on a hot summer day.

Aviators use a "standard day" as a baseline when computing aircraft performance. The standard day is defined by a sea-level barometric pressure of 29.92 inches of mercury and an air temperature of 59 degrees Fahrenheit. As the air temperature varies, so does the density of the air. This variation is used to calculate "density altitude" which is a way of assigning a value to the density of the air. For example, on a standard day at a sea-level airport such as DCA, the density altitude and the airport altitude are the same. However, a temperature of ninety degrees raises the density altitude to 2,300 feet. In other words, even though the airplane is departing from a sea-level airport, it will perform as if it is departing from an airport that is located 2,300 feet above sea-level. This will result in a longer takeoff roll and decreased climb performance. Cold temperatures have the opposite effect resulting in improved performance. Variations in barometric pressure and humidity similarly affect density altitude.

Density altitude is used to determine the maximum takeoff weight for every flight. The maximum takeoff weight is the most restrictive of the following considerations. First, every airplane model has a maximum structural weight that is determined by the aircraft manufacturer and can never be exceeded. This is not normally a limiting factor at DCA because performance limitations are more restrictive. The next consideration is the accelerate-stop distance which is the maximum weight at which an airplane can accelerate to flying speed, abort the takeoff, and safely stop on the remaining runway. This is quite often a limiting factor on the short runways at DCA but not on the long runways at IAD. Finally, a climb limiting weight must be determined which is the maximum weight at which an airplane can accelerate to flying speed, lift off, experience an engine failure, and climb above obstacles in the departure path by a specified height. This is quite often a limiting factor when departing DCA to the north due to obstacle clearance but once again, not at IAD. The long runways and straight departure paths at Washington Dulles allow for performance that exceeds the structural limit of the airplane. In other words, the most restrictive weight is the manufacturer's structural limit, but the airplane could meet the performance requirements at an even higher weight.

The takeoff weight of an airplane is a combination of the empty airplane weight plus the payload (passengers and baggage) and the fuel. The empty weight of the airplane is fixed so the only variables are payload and fuel. On average, the weight of a fully loaded airplane consists of 4% baggage, 17 % passengers, and 25% fuel. The remaining 54% is the weight of the airplane. As takeoff weight increases, so does the amount of fuel that the engines consume which in turn increases operating cost. For that reason, passenger planes are typically dispatched with enough fuel to reach their destination plus the required reserves, and nothing more. Flights that carry enough fuel to fly from DCA to the west coast cannot carry a full payload because they are limited by performance.

Modern jet engines are very reliable, so engine failures are rare. Nonetheless, pilots are trained to deal with engine failures during all phases of flight. The absolute worse time for an engine to fail is right after liftoff when the air speed is slow and the altitude is low. When a twin-engine airplane loses one of its two engines, climb performance diminishes drastically. The pilot must maintain the proper pitch attitude, within one or two degrees, or the airplane will not climb. Additionally, the pilot must maintain directional control. With full power on one side of the airplane and no power on the other, the airplane yaws (turns) in the direction of the failed engine. The pilot regains directional control using the rudder which is controlled by his feet. The rudder input will in turn cause the airplane to roll so the pilot moves the control wheel or joystick in the opposite direction to level the wings. For example, if the left engine fails, the airplane will yaw (turn) to the left. The pilot will apply right rudder to stop the turn and left control wheel or joystick to level the wings while maintaining the proper pitch attitude with very little room for error.

At most airports, like Washington Dulles, the pilot will declare an emergency and inform the controller that he has "lost an engine" and that he will be flying straight ahead until he reaches a safe altitude. This simplifies the task of stabilizing the flight path of the crippled airplane so the pilots can begin to address the engine failure. On January 8, 2026, a United Airlines Boeing 777 departed IAD for a 15-

hour nonstop flight to Japan. As the flight lifted off, one of their two engines suffered a catastrophic failure. The crew declared an emergency and informed the controller that they would be flying straight ahead. The controller asked if they would like a turn back towards the airport and the United crew replied, negative, we are climbing straight ahead. These pilots had their hands full flying a heavily loaded airplane on one engine while the failed engine was on fire. In addition to the 290 souls on board, they were loaded with nearly 40,000 gallons of fuel. To put that in perspective, those shiny tanker trucks that you see delivering fuel to your local gas station carry about 8,000 gallons. This airplane was carrying the equivalent of five tanker trucks of jet fuel. The flight eventually returned safely to IAD, but it took the pilots nearly 30 minutes to complete the appropriate checklists and dump enough fuel to reduce their landing weight, before they could put the airplane back on the runway. The full ATC audio for this flight is available on the internet by searching United 803, January 8, 2026.

When departing runway 1 at DCA, flying straight ahead is not an option. Pilots must comply with a special "engine-out" procedure to avoid prohibited airspace and maximize obstacle clearance. This procedure requires the pilot to make a left turn after liftoff to avoid flying over the White House, but not too far left or the airplane will pass over the Pentagon. A few more turns to dodge tall buildings and radio antennas then the pilots can begin to address the engine malfunction. This is slightly different than the normal takeoff procedure and it requires the non-flying pilot to go "heads down" to activate the engine-out procedure in the navigation system. This was a four-step process in the Airbus A 320 models that I last flew so the non-flying pilot must divert his attention during a critical time to accomplish this task. Once these steps are complete and the airplane is in a stabilized climb, the pilots can finally address the malfunction. Pilots are required to review and brief this procedure prior to every takeoff. Some pilots fly to DCA on a regular basis and become very familiar with this procedure, but others can fly for months and never land at this airport. Nonetheless, all these pilots are qualified in the eyes of the FAA. You can add this procedure to the list of reasons why pilots rank DCA as a challenging airport. In my opinion, procedures like this do more to justify the existence of a flight than they do to promote safety.

To reduce operating costs and maximize profits, airlines try to put as many paying passengers on a flight as possible. When they cannot squeeze any more rows of seats on a plane, they buy longer planes. It would be prohibitively expensive, and time consuming to design and build new airplanes, so manufacturers build longer versions of existing models, commonly referred to as "stretched" airplanes, due to their elongated appearance.

The Boeing 737 and Airbus A 320 family of airplanes dominate the "beyond perimeter rule" market at DCA. The original model of the 737 went into service around the same time that the first perimeter rule was established. It had a maximum structural weight of 110,000 pounds and was typically configured to carry about 100 passengers. The latest "stretched" version of the 737, which flies nonstop from DCA to San Francisco, has a maximum structural weight of 182,000 pounds and is typically configured for about 170 passengers. The

Airbus A 320 models have also been stretched. The A 321 NEO flies nonstop from DCA to Los Angeles with a maximum structural weight of 206,000 pounds and is configured for nearly 200 passengers. Neither of these airplanes can take off with a full payload from the short runway at DCA with that much fuel onboard. The payload must be reduced to reach the maximum allowable takeoff weight. In other words, these flights are "maxed out" every time they take off which means that they barely meet the takeoff requirements. Some will argue that these "beyond perimeter rule" flights are safe because they are legal, they comply with FAA limitations. I would argue that it is legal to ride a bicycle on most roads in the US but that does not make it safe. According to the Insurance Institute for Highway Safety, 1,155 bicyclists were killed in crashes with motor vehicles in 2023.

Minimum and maximum requirements should not be viewed as goals to be achieved; they are limits not to be exceeded. In my opinion, the cumulative effect of launching heavily loaded airplanes from a short runway into congested skies that are surrounded by prohibited airspace and obstacles, which require pilots to comply with a complex special procedure. is pushing the limits of safety. Personally, if I were booking a flight from Washington to the west coast, I would depart from Dulles. IAD provides a higher level of safety for long-range flights which can depart without reducing the payload and still have performance to spare. And let's not forget, IAD was built to accommodate these flights.

Regional Jets have also been stretched to accommodate more passengers which increases takeoff and landing distance. The CRJ family of airplanes has doubled its seating capacity since its introduction.

Landing performance must also be calculated to ensure that the runway is long enough to safely stop the airplane, and that the climb requirements can be met in the event of an aborted landing. The wind at the time of this accident was 320 degrees at 17 knots with gusts to 25 knots. PSA 5342 was conducting a visual approach to runway 1 which would result in a 50-degree crosswind during landing. That is a significant amount of crosswind but not extreme. The tower controller asked the CRJ crew if they could land on runway 33. This is a common request at DCA but not for the benefit of the arriving airplane. It is to help with airport congestion. In the time that it takes for an airplane to transition from runway 1 to runway 33, the tower can launch a departure on runway 1. It is the pilot's discretion to accept or reject this request. Runway 33 is considerably shorter than runway 1 but that is somewhat offset in this situation by landing with a headwind versus a crosswind. Regardless of the wind, there is another significant consideration before accepting this runway change that I will discuss in the next chapter. Runway 33 accounts for only four percent of all the landings at DCA.

When the prevailing winds blow from the south, runway 19 is the preferred runway. This runway provides the best takeoff performance because the departure procedure follows a straight path over the Potomac River with no significant obstacles. However, runway 19 has a complicated arrival procedure because prohibited airspace and obstacles require that pilots fly a convoluted path to the runway. When the ceiling and visibility are very low, approach and landing is not possible on this runway. The remaining runways are not suitable for takeoff or landing except for small airplanes or rare circumstances such as a very light weight or extreme wind conditions.

Although aircraft performance was not a direct factor in this accident, it is one of the top reasons why DCA is inadequate for its current use.

Chapter 4
Approaching the Airport

WHEN WEATHER CONDITIONS RESTRICT VISIBILITY, pilots continue to take off and land using their flight instruments. Flying an airplane soley by reference to instruments is arguably the most challenging skill set that pilots must master. Our sense of balance relies heavily on vision. When pilots lose sight of the horizon, they can easily experience spatial disorientation or vertigo. This is a powerful sensation that can trick the pilot into taking the wrong action. It is the leading cause of accidents when "non-instrument rated" pilots inadvertently fly into conditions that interfere with visual orientation. Sadly, loss of airplane control due to spatial disorientation has claimed many lives including those of John F Kennedy Jr., his wife, and sister-in-law.

JFK Jr. was not an instrument-rated pilot but, when he departed New Jersey for Martha's Vineyard on a hazy summer evening, the weather conditions met the minimum requirements for visual flight. However, once he crossed the Rhode Island shoreline, where lights from the ground were replaced by the darkness of the Atlantic Ocean, the sky and the ocean appeared as one, creating a featureless horizon which led to spatial disorientation followed by loss of airplane control.

With proper training and practice, pilots learn to ignore vertigo induced misinformation and trust their flight instruments. The cockpit voice recorder from the accident helicopter revealed that Captain Lobach experienced symptoms of vertigo prior to the collision but it is not clear if that was a factor at the time of the collision.

When airports are obscured by clouds and precipitation, pilots continue to land by following an "Instrument Approach Procedure." Instrument approach procedures fall into one of two categories, precision or non-precision. A precision approach, which is called an ILS or Instrument Landing System, provides both horizontal and vertical guidance to the touchdown point on the runway. These approaches are extremely accurate and allow pilots to land with nearly zero visibility. Runway 1 at DCA is served by an ILS approach. The instrument approach chart provides all the information that a pilot needs such as, the frequency of the navigation aid, a depiction of the approach course, minimum altitudes and instructions to follow during an aborted landing attempt. See Figure 3. I have removed some superfluous information to unclutter the chart. Take notice of the prohibited area P-56 that is outlined in hash marks. An immediate left turn is required in the event of an aborted landing to avoid this airspace. This is not exactly the ideal layout for such a busy airport. Due to restricted airspace, this is the only precision approach at DCA. By comparison, five of the six runways at Washington Dulles are served by precision approaches.

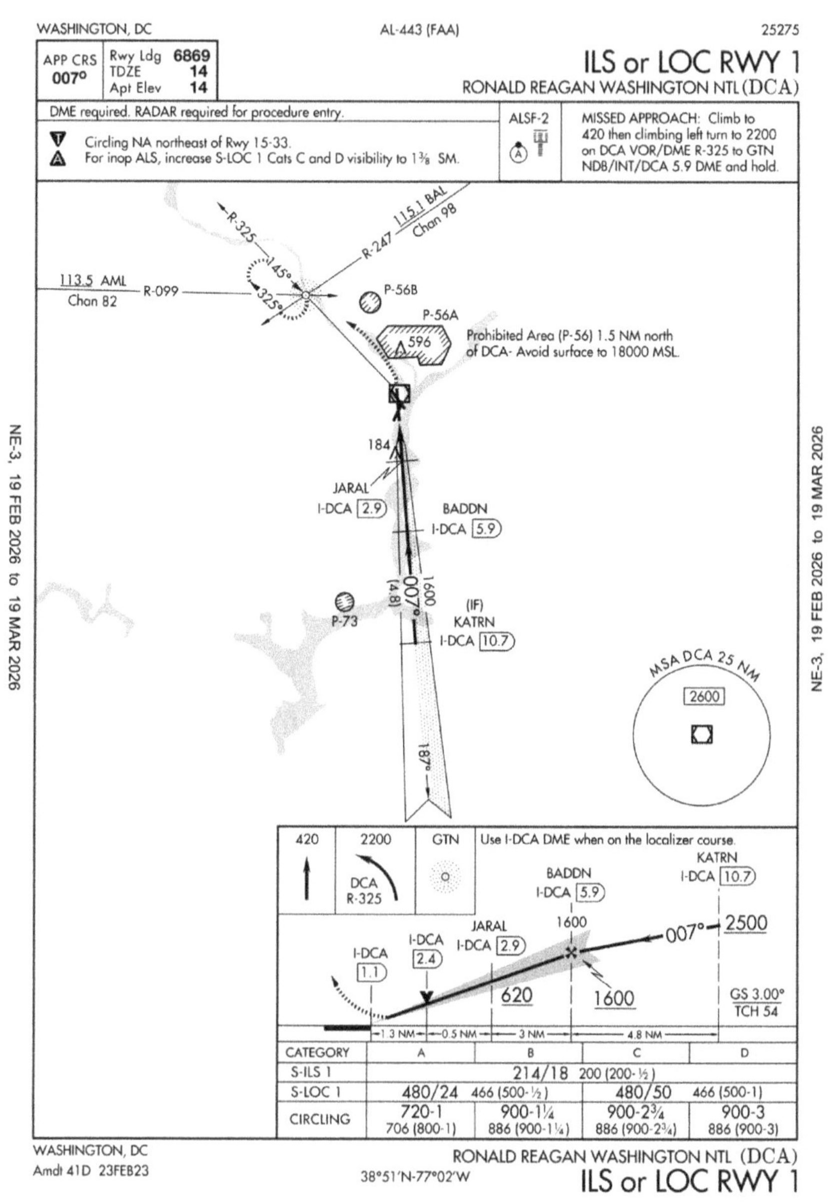

Figure 3. Precision Approach, Runway 1 at DCA. Note the prohibited area beyond the runway. Some information has been removed from the chart to reduce clutter.

Source: FAA.gov

A non-precision approach, as the name implies, is less accurate. These approaches provide horizontal guidance to the runway but not vertical guidance. In place of vertical guidance, pilots fly a series of step-down maneuvers by navigating to a designated point and then descending to a predetermined altitude before leveling off again. These approaches have a higher minimum descent altitude to compensate for their decreased accuracy. The minimum descent altitude is the lowest altitude that a pilot can descend to without having the runway in sight. If the runway is not visible, the pilot must abort the approach and follow a missed approach procedure as specified on the approach chart.

A precision approach, which requires a straight path to the runway, is not possible for runway 19 at DCA due to prohibited airspace. Instead, this runway is served by a non-precision approach which winds its way along the Potomac River to the airport. The final approach course is offset from the runway so the pilot must manually turn the airplane about forty degrees to the right, at low altitude, to align with the runway. Again, this is not exactly the ideal scenario for such a busy airport. See Figure 4.

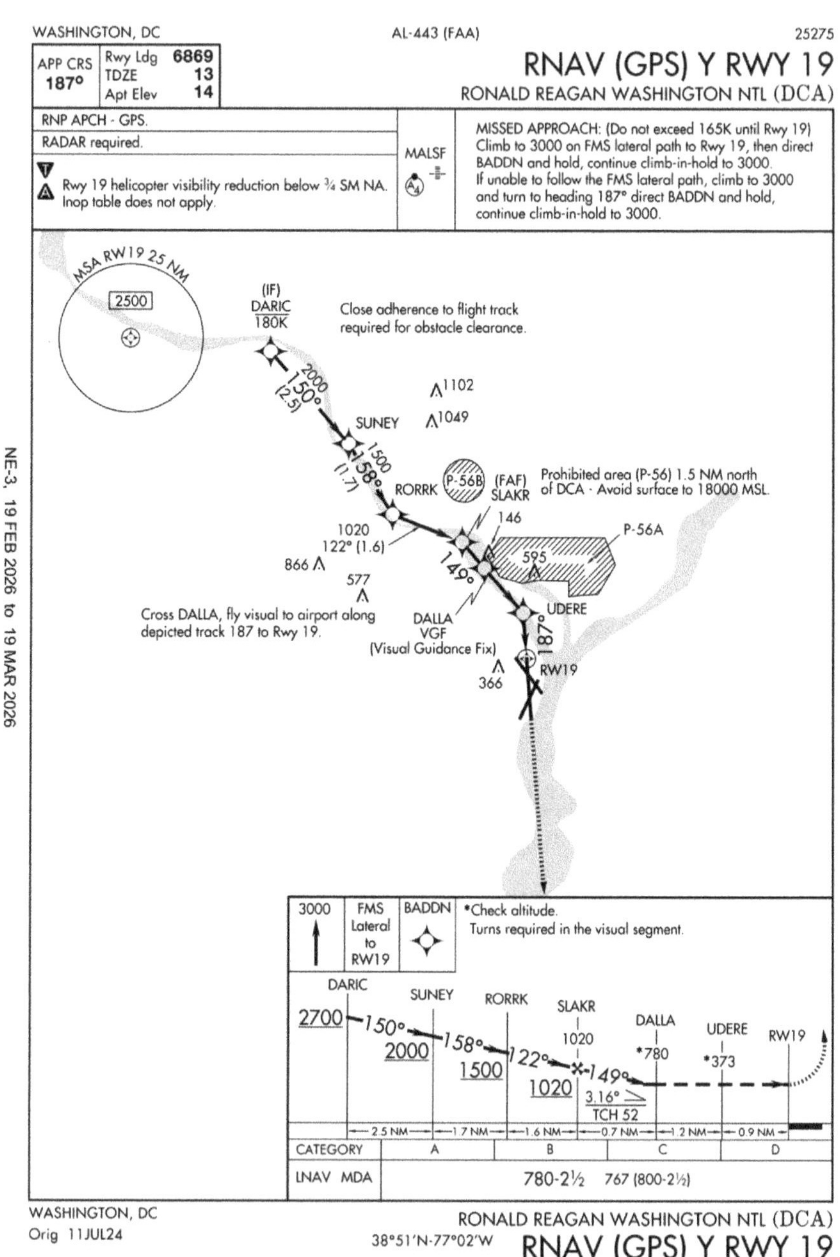

Figure 4. Non-precision approach. Note the offset course to the runway necessitating a high minimum descent altitude.

Source: FAA.gov

The QR code below provides a link to a YouTube video of an actual non-precision approach to runway 19. As you can see in Figure 5, the pilot must make a 40-degree turn at low altitude to align with the runway. This is not the ideal scenario for the nation's busiest runway but necessary due to prohibited airspace near the airport.

Figure 5. To view the video, scan the QR code or type the web address into your browser.

Source: Ben Lochary

https://tinyurl.com/rnav19

When the weather is clear, pilots can fly a visual approach in lieu of an instrument approach. This allows traffic to flow faster because airplanes can be spaced closer together. As the name implies, a visual approach is flown by looking through the windshield in the same way that we drive cars. In most cases visual approaches are flown by "the-seat-of-the-pants" meaning that the pilot uses his eyesight and judgement to navigate to the runway. Flying a visual approach without vertical guidance can be deceiving. On a clear night airports appear closer than they are, and this can lead a pilot to descend too soon. During June of 2024, a Southwest Airlines Boeing 737 crew descended dangerously close to the ground during a night-time visual approach to Oklahoma City setting off the low altitude alert in the control tower. As a rule of thumb, pilots allow 3 miles for every 1,000 feet of descent. This flight was 9 miles from the airport so it should have been at 3,000 feet but for some reason the crew descended to 500 feet before an alert air traffic controller intervened, quite possibly averting a catastrophe.

Some airports have "charted" visual approach procedures which dictate that pilots fly a specific course to the runway. DCA has two of these approaches. One for landing north on runway 1 and one to the south for runway 19. This is like an instrument approach except that the runway must be in sight before commencing the approach. On the night of the DCA accident, Captain Campos and First Officer Lilley were flying a charted visual approach to runway 1 called the Mount Vernon Visual Approach. See Figure 6. This is the easiest way to approach DCA. ATC vectors you until you are over the Potomac River heading almost directly towards the airport. Once cleared for the approach, the pilot will intercept the ILS approach and follow it to the runway. Technically this is a visual approach, but it can be flown with the same accuracy as a precision approach.

DCA is a very busy airport with arrivals and departures sharing a single runway. There is often a line of airplanes waiting for takeoff and runway 1 is the only suitable choice for north operations. Once an airplane touches down, the local controller will clear a waiting airplane into position on the runway in preparation for takeoff. As soon as the airplane that just landed clears the runway, the departing airplane is cleared for takeoff. Quite often there is another airplane on final approach for landing. If that airplane reaches the runway before the departing airplane becomes airborne, the arriving airplane will be forced to abort the landing. Sometimes ATC will ask the arriving airplane if they can land on runway 33. This is not for the benefit of the airplane that is landing but to create additional space between the departing and arriving airplanes. This is a common request at DCA due to the high volume of traffic that is sharing a single runway.

Captain Campos accepted this runway change. The proceeding CRJ did not and landed straight ahead on runway 1. I am not implying that Captain Campos made a mistake by accepting the change because it is a judgement call. Many factors must be weighed in making this decision, such as wind, landing weight, prior experience with this procedure, and pilot fatigue to name a few. A disadvantage of changing to runway 33 is that there is no guidance to this runway other than looking through the windshield and trusting your judgment which requires the pilot's full attention.

As a result of lessons that were learned from past landing accidents, pilots are taught to have the airplane stabilized by 1,000 feet above the airport. Under ideal conditions this can be delayed to 500 feet. Stabilized means that the airplane should be on the proper course and descent path, in landing configuration, the engines above idle and airspeed no more than 10 knots above the target speed. If you don't meet these requirements, you must abort the landing.

To transition from runway 1 to runway 33 the pilot must manually fly the airplane by the "seat of his pants" without the aid of the autopilot or electronic guidance. First, he will turn the airplane about thirty-degrees to the right for a short time and then back to the left to line up with runway 33 while continuing the descent to touchdown. He must visually judge his vertical path so that he is not too high or too low as he approaches the runway. There is a visual aid adjacent to this runway to help the pilot stay on the proper vertical path. It is called a Precision Approach Path Indicator or PAPI and it consists of red and white lights. If the pilot sees all white lights, he is too high. All red lights indicate too low. A combination of red and white lights indicate that the pilot is on the correct vertical path. All of this happens below 1,000 feet and demands the pilot's full attention.

As if that is not enough, dozens of helicopter flights traverse the skies around DCA every day. By some accounts, there are as many as 28 helicopter operators crisscrossing the area on a regular basis. These include military, as well as multiple medevac, and law enforcement operations. To organize the flow of traffic, designated helicopter routes surround the airport. One of these routes, helicopter Route 4, crosses the approach path to runway 33 less than one mile from the end of the runway. If a helicopter following this route flies no higher than the charted altitude of 200 feet, and the approaching airplane is on the proper visual descent path, the two aircraft will have no more than 75 feet of separation. Normal vertical separation for aircraft in level flight is 1,000 feet. In visual conditions this separation can be reduced to 500 feet. Allowing 75 feet of separation between two aircraft, one of which is descending, is perilously unsafe, yet this practice has been in place for years.

Captain Campos guided his CRJ along the proper descent path throughout the approach, but the helicopter was nearly 100 feet above the charted altitude. The reason for the altitude deviation, as well as alarming details about this deadly helicopter route, were revealed at a public hearing that was conducted by the NTSB. I will explain the details from that hearing later in this book, but before I do, more background information is necessary for you to understand the information that the hearing revealed.

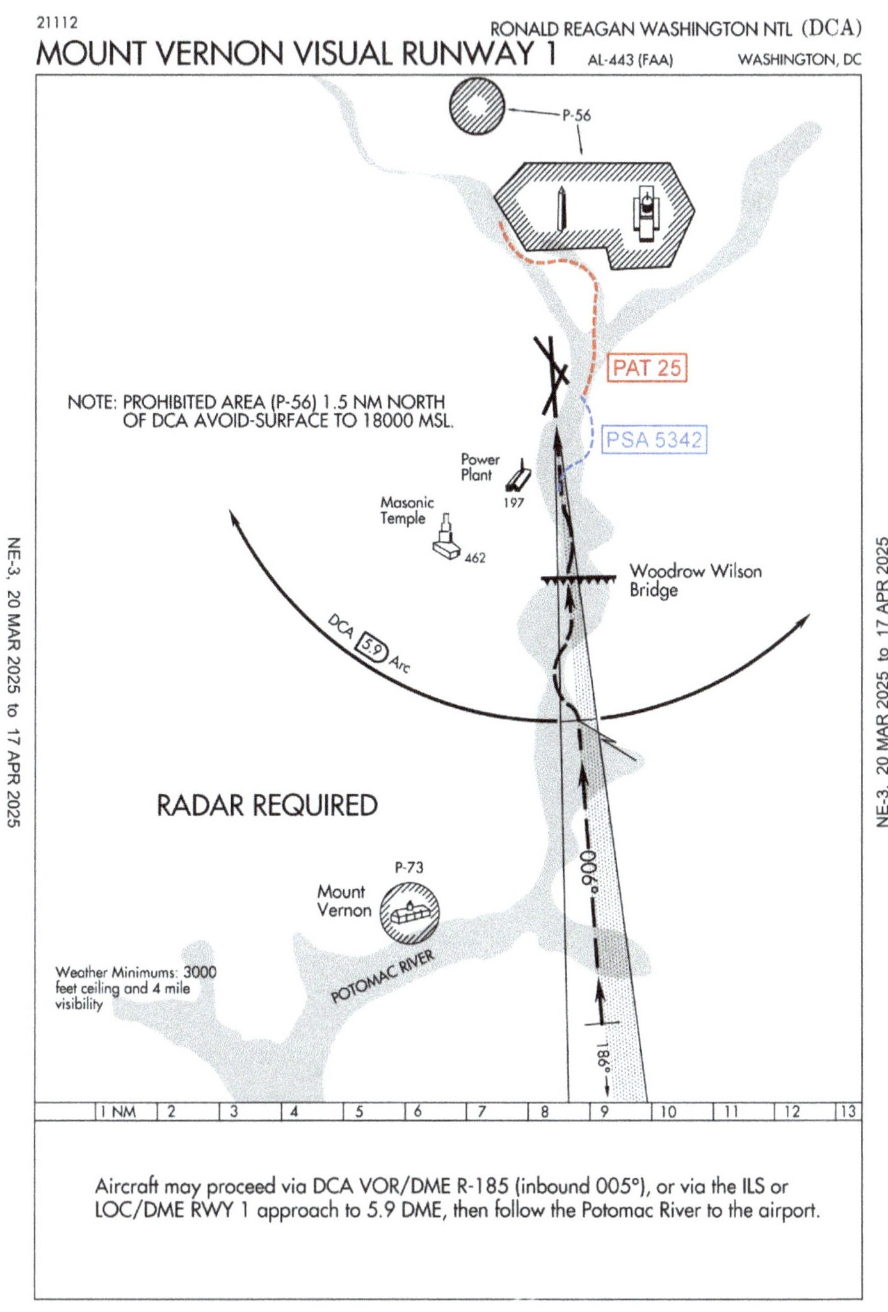

Figure 6. Visual approach to runway 1.

Source: FAA.gov

Figure 7. Diagram from the NTSB preliminary accident report.
Source: NTSB preliminary accident report.

This diagram was taken from the NTSB preliminary accident report. The blue line shows the path of the CRJ (5342) as it transitioned from runway 1 to runway 33. The red line shows the path of the helicopter (PAT 25). Looking at this diagram, it is difficult to understand how this is an approved procedure.

Chapter 5
Midair Collision Accidents

SADLY, THE JANUARY 29 COLLISION AT DCA is not the first time that a civilian passenger plane was taken down by a military aircraft on a training flight. In 1958 an Air Force fighter jet collided with a United Airlines passenger plane near Las Vegas leaving no survivors. The United flight was flying on a designated route from Los Angeles to Denver while the military flight was conducting training near that same route. The United flight was under the control of ATC (Air Traffic Control) but not in radar contact. In those days radar tracking was only available near large, busy airports. In lieu of radar, ATC relied on position reports. Pilots would make radio contact with air traffic controllers and report the time that they crossed a designated reporting point and the estimated time that they would arrive at the next reporting point. Controllers would then plot these positions on a map. A similar, but more sophisticated, system is still used today where radar coverage is not available.

The military flight was flying under Visual Flight Rules (VFR) and it was their responsibility to see and avoid other traffic. The pilot of the military jet could not see outside because he was wearing a "view limiting device" to simulate flying in the clouds. It was up to the instructor pilot to watch for traffic, but the design of the cockpit limited his view. The two aircraft could not hear each other because they were on different radio frequencies. The United flight communicated with ATC on one frequency and the Air Force jet communicated with a military controller on a different frequency. There was also no communication between the two controllers. In those days military traffic was controlled separately from civilian traffic even though both occupied the same airspace.

Captain Lobach and Chief Warrant Officer Eaves were conducting a training flight under Visual Flight Rules (VFR) and they accepted responsibility to maintain visual separation from the CRJ. As part of their recurrent training mission, they were wearing night vision goggles which limited their peripheral vision. The DCA controller communicated with the helicopter and the CRJ crews on separate radio frequencies. The controller alerted the helicopter crew about the presence of the CRJ but never alerted the CRJ crew about the presence of the helicopter. If Captain Campos and First Officer Lilley had been warned of the intruding helicopter, it is likely that they could have maneuvered their airplane to avoid the collision.

Just one month after the Las Vegas accident, another civilian passenger flight was taken down by a military plane that was on a training flight. An Air Force jet trainer ran into a Capital Airlines flight that was enroute from Pittsburgh to Baltimore and there were no survivors. Once again, the military jet was on a training flight under Visual Flight Rules (VFR) and it was their responsibility to see and avoid other traffic. The practice of allowing military aircraft to conduct training flights in airspace that is used for scheduled passenger flights is unacceptable, yet this practice continues today even though there are designated military training areas across the contiguous United States where civilian traffic is prohibited.

These accidents prompted Congress to pass, and President Eisenhower to sign into law, the Aviation Act of 1958. This act abolished the Civil Aviation Agency (CAA) and established the Federal Aviation Agency (FAA), later renamed the Federal Aviation Administration. The FAA was tasked with regulating safety in the airline industry and managing airspace for both civilian and military flights. Since then, all air traffic in the United States is controlled by ATC except for military airports where civilian airplanes are not allowed except by special arrangement or during an emergency.

In 1978 a tragic midair collision occurred over San Diego killing 144 people and this accident also had similarities to the DCA midair collision. A Boeing 727, operated by the original Pacific Southwest Airline, was flying a visual approach when it collided with a small private airplane that was conducting a training flight. The airline crew initially reported that they had the traffic in sight and would maintain visual separation. When a pilot accepts a clearance to "maintain visual separation" he accepts responsibility to avoid that traffic. If he subsequently loses sight of the traffic, he must notify ATC immediately. If a pilot is not one-hundred percent certain that he has the correct traffic in sight, and can keep the traffic in sight, he should not accept visual separation in which case it is the controller's responsibility to positively separate the two aircraft.

Just prior to the San Diego collision, the controller received an automated conflict alert indicating that the two airplanes were on a potential collision course. It is possible that two airplanes can safely pass inside of the parameters that trigger this alert provided that one aircraft clearly sees and avoids the other. In this case the controller took no action because the airline crew had already accepted responsibility for avoiding the small airplane. The collision occurred a few seconds later. The NTSB accident investigation revealed, among other things, that ATC had the capability to positively separate these airplanes but relied on the pilots to maintain visual separation despite the conflict alert.

The DCA collision has too many similarities to the San Diego collision. Most importantly, the automated traffic alert that was triggered in the control tower indicating that the CRJ and the helicopter were on a potential collision course. The controller responded by asking the helicopter crew if they still had the CRJ in sight and they replied that they did and would maintain visual separation. The collision occurred a few seconds later. The controller had the ability to positively separate these two aircraft but instead relied on visual separation. Visual separation is a useful tool for managing air traffic, but it has limitations. As we learned from the San Diego and the Washington collisions, an automated conflict alert requires additional action by the controller.

Due to the high volume of helicopter traffic near DCA, a second local control position was established to manage only helicopter traffic. This is testament to the hazardous nature of allowing transient helicopter flights near a busy airport. The helicopter local control position was not staffed at the time of this tragic accident. If the second local controller had been on duty, to focus his attention solely on PAT 25, this traffic conflict may have been handled differently. In my opinion, the practice of allowing helicopters to fly dangerously close to passenger planes is unacceptable and allowing one controller to do the job of two does not prioritize safety. The lone local controller was, through no fault of his own, overloaded. It has been widely reported that the current shortage of controllers has led to six-day work weeks and fatigue.

The helicopter crew had an airplane in sight, but it was not PSA 5342. The airplane that they were focused on was straight ahead, and PSA 5342 was at their ten o'clock position. The helicopter crew was conducting pilot proficiency training, and night vision goggle proficiency training. Night vision goggles allow pilots to see things in the dark, that they would not otherwise see. but they restrict peripheral vision. Imagine yourself looking through a pair of binoculars. You can see straight ahead quite well but not to the side. This explains why the helicopter crew did not see the CRJ even though it was not far away, but the bigger question is why helicopters are flying so close to passenger planes, especially while training.

FAA statistics reveal that between 1987 and 2021 there were over 2,000 near midair collisions involving military aircraft and 258 of those involved a conflict between a military and a commercial aircraft. Of these, 34 were considered critically close. Occurrences like this should spark immediate change but that is rarely the case.

At the time of the fatal Air Florida crash at DCA in 1982, which was due in part to improper deicing procedures, there were no specific guidelines for deicing airplanes. There was simply a regulation, commonly referred to as the clean airplane concept, that prohibited pilots from taking off with frozen precipitation adhering to critical surfaces of the airplane. Following this accident the FAA re-emphasized the importance of complying with this rule. Almost five years later, a very similar fatal accident occurred at Denver's Stapleton International Airport but, once again, no significant changes were made to deicing procedures.

Twenty-eight months later another nearly identical accident occurred at the Dryden Regional Airport in Ontario, Canada. Once again, a pilot attempted to take off with snow adhering to the airplane which resulted in a crash immediately after liftoff and loss of life. This accident triggered a lengthy judicial inquiry headed by the Honorable Virgil P. Moshansky. Although the pilot was clearly at fault for attempting to take off with contaminated wings, it was also determined that competitive pressures negatively influenced safety standards, and the commission admonished the airline industry for putting profit ahead of safety.

Judge Moshansky stated that "aviation safety should not be looked on as merely a selling point or marketing device, nor should it be viewed as some abstract goal by which to satisfy the minimum standard required by the regulator…aviation safety must be viewed from management on down, as an obligation of trust to the traveling public…."

Just three years after the Dryden accident the same scenario was repeated at New York's Laguardia airport and in the same type of airplane. This flight was operated by my employer, US Air. Once again, the pilot attempted to take off with snow accumulation on the airplane resulting in a crash after liftoff and loss of life. Upon hearing the details of this accident, Judge Moshansky stated, "My God, its Dryden all over again." He also said that the LaGuardia accident could have been avoided if warnings in the report on the Dryden crash had been heeded.

That report had been shared with the FAA and the US airline industry. Recommendations in the report included applying anti-icing fluids after deicing, deicing airplanes close to departure runways to shorten the effects of ground delays, and training for ground and flight crews on deicing procedures. Today, these recommendations are standard operating procedure throughout the airline industry. To my knowledge there has not been an accident of this type since.

It is very disturbing that so many lives were lost before proper action was taken. How many collisions and close calls will it take before necessary changes are implemented at DCA. Some members of Congress have called for reduced flying by the Pentagon in the vicinity of the airport, the creation of an alternative route for helicopters, and a review of existing airline traffic volume.

The helicopter route that led to this accident was permanently closed following alarming statistics that were uncovered by the NTSB during the investigation. That is a much-needed step in the right direction, but more must be done including a review of airline traffic volume at DCA. Arriving and departing airplanes should be adequately separated so that pilots and controllers are never put in a position where they must push the limits of safety as a daily course of action. The public hearing, that I will cover in Chapter 8, exposed a loophole used by American Airlines to enhance its schedule which contributes to the traffic congestion that plagues DCA every day. But first, I need to explain collision avoidance technology and air traffic control procedures which are central to the investigation.

Chapter 6
Traffic Collision Avoidance System

TO GUARD AGAINST MIDAIR COLLISIONS, a system was developed to warn pilots of potential traffic conflicts. It is called the Traffic Collision Avoidance System or TCAS and it uses airplane transponders to monitor surrounding traffic. A transponder is an electronic device, installed on airplanes, that responds to ATC radar. An airplane that is not equipped with a transponder appears as an anonymous dot or blip on the controller's radar display. Transponder equipped airplanes appear with identifying information such as flight number, type of airplane and altitude, allowing the controller to positively identify every aircraft that appears on their radar display. TCAS has been required equipment on passenger airplanes for over thirty years, however, it is not required for most helicopter operations.

If two TCAS-equipped airplanes appear to be on a collision course, the system will issue a Traffic Advisory or TA to both airplanes. The TA consists of an aural "TRAFFIC -TRAFFIC" alert and a visual display in the cockpit indicating the location and relative altitude of the conflicting traffic. Pilots are trained to respond to a TA by attempting to visually acquire the traffic while preparing to take evasive action. If evasive action is necessary, the TCAS will generate a Resolution Advisory or RA. The RA consists of an aural command such as, "CLIMB-CLIMB" OR "DESCEND-DESCEND" with corresponding indications on the pilot's flight instruments. This is a coordinated command, that is, one airplane is commanded to climb and the other to descend.

In my opinion, TCAS is one of the greatest safety innovations of modern aviation. Ask any experienced pilot and they will tell you that they have responded to a Resolution Advisory at least once. I responded to several during my career and any one of those events may have resulted in a mid-air collision if not for TCAS. Many pilots, me included, were skeptical of TCAS when it was first introduced. A basic rule of instrument flying is that pilots must maintain their assigned altitude unless cleared otherwise by ATC. TCAS Resolution Advisories require the pilot to respond in five seconds or less, which leaves no time to notify ATC that you are leaving your assigned altitude until after the fact. This was a big adjustment for both pilots and controllers. Another concern involved trusting the system. TCAS is designed to command one airplane to climb and the other to descend, but what if it gives the same command to both aircraft? These concerns might seem crazy by today's standards, but this was long before the "gee whiz" technology that we take for granted today. Our smart phones have better navigation capability than passenger planes did back then.

The next generation of TCAS is on the horizon but the current version is still proving its worth. As I write this, a Southwest Airlines 737 en route from California to Nevada took evasive action in response to a TCAS RA quite possibly averting a disaster. Two flight attendants were injured, and the passengers had the scare of a lifetime, but the airplane landed safely.

Unfortunately, the TCAS Resolution Advisory could not prevent the DCA collision because it is inhibited at low altitudes to prevent nuisance alerts during takeoff and landing. Although the Resolution Advisory function is inhibited, the Traffic Advisory function remains operational. The CRJ crew received a Traffic Advisory just seconds before the Black Hawk helicopter ran into them. At that point Captain Campos and First Officer Liley would have immediately looked to their right to locate the offending traffic which was camouflaged against a background of bright lights. Spotting traffic under those conditions is difficult at best. Without visual confirmation of the traffic, the CRJ crew would not know what evasive action to take.

PAT 25 was not equipped with TCAS. If it had been, the helicopter pilots would have received a Traffic Alert indicating that the CRJ was about to pass in front of them from left to right. The investigation revealed that seven of the 12th Aviation Black Hawk helicopters were equipped with TCAS, but Army pilots were not trained to use it. As a result of this accident, the Army is training its pilots to use this valuable safety tool.

Although PAT 25 was not equipped with TCAS, it was equipped with a relatively new technology called ADS-B or Automatic Dependent Surveillance Broadcast. There are two components to this system, ADS-B Out and ADS-B In. Without getting too technical, ADS-B Out equipped aircraft broadcast their GPS position directly to aircraft that are equipped with ADS-B In, as well as “radar like” displays used by ATC. This system is more precise than radar and it has additional advantages. Radar is limited by line-of-sight and distance while ADS-B is not. ADS-B updates aircraft position every second while radar takes four to six seconds to update. Eventually ADS-B will replace radar because it has fewer limitations.

ADS-B Out is required for all aircraft that operate in Class B airspace such as that which surrounds DCA, however, military aircraft are exempt from turning it on to prevent unwanted tracking of critical missions. Fight tracking apps, such as FlightAware, derive their tracking information from ADS-B and other sources. The Army asserts that training flights, like that flown by PAT 25, are critical because they combine practice landings at sensitive locations that should remain anonymous. It seems reasonable to me that the military should be free to conduct critical missions without broadcasting their position to the world but training flights like this should not be allowed when the airspace is filled with passenger planes.

PAT 25 was not transmitting ADS-B data at the time of this accident. In fact, the investigation revealed that this helicopter had not transmitted ADS-B data for the past 730 days. Investigators discovered that even when the system was turned on, multiple Black Hawk helicopters from the 12th Aviation Battalion were not transmitting this data due to a programming error when the system was installed. That problem has since been fixed and the Army is now performing regular checks to verify that the system is transmitting properly.

Although ADS-B Out is required, ADS-B In is not. If PAT 25 had been equipped with ADS-B In, Captain Lobach and Chief Warrant Officer Eaves would have had a cockpit display, similar to a TCAS display, showing that the CRJ was about to cross in front of them from left to right. Few Army helicopters are equipped with ADS-B In due to cost and the fact that it is not required. As a result of this accident the Army has authorized the purchase of 1,600 ADS-B In systems for its helicopter fleet.

According to the Associated Press, Brigadier General Matthew Braman, the Army's head of aviation, stated that "there was no point during the flight where the jet and the airport control tower could not see the Blackhawk." I respectfully disagree with the General's assertion that the CRJ crew should have seen the helicopter. The controller clearly saw that the helicopter was on a potential collision course with the CRJ. He alerted the helicopter crew on two occasions about the presence of the CRJ but never alerted the CRJ crew about the presence of the helicopter. Although the helicopter would have been visible on the CRJ's TCAS display, along with every other aircraft that was in the traffic pattern at that time, Captain Campos and First Officer Liley had no reason to anticipate a traffic conflict so close to touchdown. They would be more concerned about potential traffic conflicts on the runway, such as a vehicle or airplane. The first warning that Captain Campos and First Officer Liley received about a potential collision was the TCAS Traffic Advisory seconds before impact. Captain Campos tried in vain to maneuver his airplane away from the intruding helicopter, but it was too late. The collision happened so fast that the innocent souls on the CRJ did not have time to panic.

ADS-B is so effective that the FAA encourages airport operators to equip ground vehicles with ADS-B Out transmitters to reduce the risk of conflict between airplanes and ground vehicles. During a recent incident at DCA, a potential collision between an airplane and a ground vehicle was averted thanks to an alert flight crew. Republic Airlines Flight 4528, a CRJ operating under contract for American Airlines, departed Detroit in the early morning hours of September 9, 2025. Upon arrival at DCA, flight 4528 was cleared for the Mount Vernon Visual Approach to runway 1. The local controller asked the crew if they could switch to runway 33 to accommodate a departure on runway 1 and the crew accepted this request. This is the same path flown by Captain Campos and First Officer Lilley but with two significant differences. The dangerous helicopter route that crossed the final approach path for runway 33 had since been shut down by order of the Secretary of Transportation as recommended by the National Transportation Safety Board. The other significant difference was the absence of darkness. As the Republic flight turned to line up with runway 33, the morning sun illuminated the airfield like a giant spotlight. Seconds before touchdown the alert Republic Airlines pilot asked the controller, "is there a vehicle on 33?" The controller, who was facing east towards the blinding sun responded, "affirmative, ahh, go around turn left heading two-five-zero." This incident could have resulted in a serious accident highlighting the value of equipping ground vehicles with ADS-B Out, and mandating ADS-B In for all aircraft.

The NTSB has long championed the use of technology to enhance safety. They have been the driving force behind countless safety enhancement such as three-point seat belts and air bags for motor vehicles, leak detection and automatic remote shut-offs for onshore pipelines, Positive Train Control which uses wireless GPS communication technology to prevent train accidents, and traffic collision and ground proximity warning systems in aviation, to name just a few. Although safety enhancements like these save lives, they are not readily accepted by industry where mandated safety requirements reduce profit. I am reminded of a conversation that I had with an FAA official years ago. I asked him why an airport that was under his jurisdiction had not adopted a simple improvement that was standard at most airports. He said that he was actively lobbying for it and would continue to do so but the airport authority would rather spend money updating the food court than spending it on something that passengers cannot see. The NTSB is actively lobbying for mandated ADS-B In, which is a no-brainer but, as recently as 2023, the FAA stated that the current requirements adequately address aviation safety and they will not pursue any additional ADS-B requirements. Requiring ADS-B Out and not requiring ADS-B In is akin to requiring cars to have left turn signals, but not right turn signals.

The preliminary NTSB investigation into this accident revealed alarming statistics. Between 2011 and 2024, an average of one TCAS Resolution Advisory per month was triggered at DCA involving a helicopter and a passenger airplane. In over half of these instances the helicopter may have been above the charted route altitude. Two-thirds of the events occurred at night. This raises the questions, who had access to these statistics and why was there no corrective action? Two days after this avoidable and tragic accident, the FAA issued a notice prohibiting helicopter traffic over the Potomac River in the vicinity of DCA. An exception was made for life saving medical, active law enforcement, air defense, and presidential transport missions. If helicopter flights near the airport are necessary, passenger flights will be delayed until the helicopter traffic has cleared the area.

On March 1, 2025, just one month after the midair collision that claimed sixty-seven lives, several inbound flights to DCA received Traffic Alerts and some were forced to abort their approach to landing. When questioned about the traffic ATC responded that there was no known traffic in the vicinity. The FAA vowed to investigate this incident. According to Senator Ted Cruz, Chairman of the US Senate Committee on Commerce, Science, and Transportation, these alerts were caused by the Secret Service and the Navy testing counter drone technology near DCA using the same frequency that the TCAS system utilizes. The Secret Service denied Senator Cruz's claims, adding: "The Secret Service did not conduct any drone system testing in the Nation's Capital Region on March 1, 2025. The agency has been coordinating with the FAA to ensure our systems do not interfere with the FAA frequencies or commercial air traffic operations." This begs the question, what systems are they using that require coordination with the FAA? For now, this remains a mystery.

One month later a Delta Airlines flight carrying 131 passengers was forced to take evasive action in response to a TCAS Resolution Advisory shortly after takeoff from DCA. The conflicting traffic was a US Air Force T-38 trainer enroute to Arlington National Cemetery for a ceremonial flyover. Flights like this are prearranged and coordinated with the DCA control tower. At 3:13 p.m., the controller in charge at DCA received instructions from the military to cease operations at 3:17 p.m. to accommodate the flyover. The Delta flight was cleared for takeoff at 3:15 p.m. and the close encounter occurred at 3:16 p.m. This incident highlights just how overburdened the airspace around DCA is and the ongoing conflict between military and civilian use of this airspace.

The high number of hourly flights at DCA combined with an antiquated runway layout leaves no room for error. To illustrate, consider the traffic congestion leading up to the tragic collision between the CRJ and the helicopter. As Captain Campos maneuvered his CRJ towards runway 33, the preceding CRJ was touching down on runway 1. As soon as that airplane touched down another airplane was cleared onto runway 1 in preparation for takeoff. When the airplane that just landed exited the runway, the waiting airplane was "cleared for immediate takeoff, no delay." The departing airplane had to accelerate past the intersection of runway 33 before Captain Campos touched down.

If the departing airplane did not clear the intersection in time, Captain Campos would be forced to abort his landing, so the controller would have been watching the progress of these two flights very closely. Around this time the traffic conflict alert was sounded in the control tower indicating that the helicopter and the CRJ were on a potential collision course. For the second time the controller asked the helicopter crew if they had the CRJ in sight. Chief Warrant Officer Eaves responded that they did and would maintain visual separation. The controller was then distracted by a radio transmission from the next arriving airplane and that is when the collision occurred. Pilots and controllers have been forced into a position where they must use every trick in the book to compensate for the overcrowding of DCA that airlines and legislators have created.

This same sense of urgency may have played a role in the deadly Air Florida crash during the winter of 1982. I remember that day very clearly. I was flying with First Officer John Wade who hails from a family of accomplished aviators, and it was reassuring to fly with a pilot of his caliber on such a challenging weather day. We landed at DCA in the early morning and were greeted by low visibility and heavy snowfall. We returned later that day to the same weather conditions. As we stabilized our plane on the instrument approach to runway 1, which at that time was identified as runway 36, we were dumbfounded when the tower controller said to the preceding airplane, when you land, I want you to roll to the end of the runway and let me know what you see, we are missing an airplane. John and I looked at each other in disbelief but without saying a word we refocused our attention on the approach, fully expecting the controller to issue go-around instructions. In aviation jargon an aborted landing is called a "go around" and it is a common event at DCA due to the close spacing of aircraft to a single runway. With just seconds to spare, we were cleared to land, then the airport was closed until the next day.

A series of events led to the Air Florida 90 crash, but I want to focus on just one. The captain failed to turn on a critical engine anti-icing system which led to erroneous engine indications. When the engine instruments indicated that the engines were producing full power, they were producing far less. As the Boeing 737 accelerated down the runway the first officer questioned the engine indications four times, but the captain dismissed his concerns. The 737 used every inch of runway to reach flying speed. Once airborne, the airplane climbed to about 300 feet, stalled, and crashed onto the 14th Street bridge before plunging into the frozen Potomac River.

Why did the captain continue when he could easily abort the takeoff? No one can say for certain, but I believe that the ATC transcript from that fateful day may provide a clue. While Air Florida 90, which is identified by the call sign Palm 90, was waiting on the runway for takeoff, an Eastern Airlines flight was on final approach to land on that same runway. The following is a portion of the control tower transmissions leading up to this accident:

– "Palm 90 taxi into position and hold, be ready for an immediate."

– "Eastern 1451, keep it at reduced speed, traffic's going to depart…."

– "Palm 90 cleared for takeoff. No delay on departure if you will, traffic's two and one-half out for the runway."

The Eastern flight touched down while the Air Florida jet was still on the runway. Separation between the two planes was less than 4,000 feet and in violation of Air Traffic Control rules. The tower could not see either airplane and the two airplanes could not see each other because the visibility was restricted by falling snow. Fortunately, John and I could not see the accident, but we spent the rest of the day in shock as we watched the news reports from the airline break room.

Is it possible that the Air Florida captain was preoccupied by the rapidly approaching Eastern flight causing him to minimize the first officer's concerns? We will never know for certain but there is a sense of urgency at DCA, which is fueled by the close spacing of arriving and departing traffic to a single runway, that can negatively influence decision making.

On May 2, 2025, just one week after Priority Air Transport flights were resumed to the Pentagon, an alert air traffic controller commanded two arriving passenger planes to abort their approach due to the proximity of an Army helicopter. According to NBC news, the helicopter was rehearsing a Joint Emergency Evacuation Plan. A Senate Commerce Committee hearing revealed that a hotline between the Pentagon's Army heliport and the DCA tower has been out of service for over two years due to construction at the Pentagon, but communication was still available through conventional "dial-up" phone lines. The FAA's deputy chief operating officer said the agency didn't know at the time of the accident that the hotline wasn't functioning. "We were not aware, but we became aware after the event, and now that we became aware of the event, we're insisting that the line be fixed before we resume any operations out of the Pentagon." It does not appear that this inoperative hot line played a role in the tragic midair collision, but it is one more "red flag" that gives cause for concern.

Chapter 7
Air Traffic Control

IN THE AFTERMATH OF THIS HORRIFIC ACCIDENT, it has been widely reported that air traffic controllers are understaffed, overworked, and hampered by outdated equipment. This is not a new problem. The Professional Air Traffic Controllers Organization (PATCO) was formed in 1968 to address these same issues. Frustrated by a lack of progress during contract negotiations, PATCO orchestrated a controller "sickout" which disrupted air traffic. Federal courts intervened and ordered the controllers back to work, but this action brought the union and the government back to the bargaining table. As a result, Congress moved ahead with modernizing equipment, hiring additional controllers, and increasing salaries.

Long before Ronald Reagan was elected as the fortieth President of the United States, and before he became the thirty-third Governor of California, he served as president of the Screen Actors Guild. In 1980, PATCO, as well as some larger trade unions, backed fellow union member Ronald Reagan in his successful run for president. Around this time, Inflation soared to double digits, which led to a demand for higher wages sending PATCO back to the bargaining table. Following seven months of unsuccessful contract negotiations, the union went on strike, which was in violation of federal law. President Reagan gave the striking controllers forty-eight hours to return to work and issued the following statement.

"Let me make one thing plain. I respect the right of workers in the private sector to strike. Indeed, as president of my own union, I led the first strike ever called by that union. I guess I'm the first one to ever hold this office who is a lifetime member of an AFL-CIO union. But we cannot compare labor management relations in the private sector with government. Government cannot close down the assembly line. it has to provide without interruption the protective services which are government's reason for being. It was in recognition of this that the Congress passed a law forbidding strikes by government employees against the public safety."

At the end of that forty-eight-hour period, President Reagan fired more than 11,000 air traffic controllers who refused to cross the picket line. Not only were they fired, but they were also banned from ever holding any government job in the future. That left a handful of non-striking controllers, supervisors, and 900 military controllers to rebuild the air traffic control system. The government projected that it would take two years for the system to return to normal staffing levels, but it took nearly a decade.

Replacing the fired controllers had long lasting ramifications. It would take years to hire and train new controllers and even more years for them to gain experience. Twenty-five years into the future these controllers would "age out" within a few years of each other and the mass hiring cycle would start again. According to the FAA's Air Traffic Control Work Force plan for the 2021–2030-time frame: "As the new controllers hired en mass in the early 1980s achieved full certification, the subsequent need for new hires dropped significantly from 1993 to 2006. This caused training percentages to reach unusually low levels. The FAA's current hiring plans will return trainee percentages to their historic averages."

As I write this, there is a nationwide shortage of 3,000 controllers. The Ronald Reagan Washington National Airport is operating at 63% of its target staffing level. Other major airports in the northeast corridor are operating at 60% to 65% levels. In 2023 only 23 of 313 FAA facilities met ATC staffing levels. The controller's union reports that 41% of its members are working six days a week and 10-hour days. Although the FAA reached its goal of hiring 1,800 new controllers, the workforce grew by only 36 after factoring in retirements. Controllers can retire with full benefits at age 50 after 20 years of service and retirement is mandatory at age 56.

Filling these vacancies is a slow process. Applicants must be less than thirty-one years of age. All initial training is conducted at the FAA Academy in Oklahoma City for two to five months. It is reported that the washout rate at the academy can be as high as 35%. Those who graduate from the academy are assigned to an FAA facility for on-the-job-training which takes one to three years. The overall washout rate can reach 50%.

The National Air Traffic Controllers Association (NATCA), which currently represents approximately 70% of the controllers in the US, has been raising concerns for years about the controller shortage. It warned Congress in 2015 that the shortage was reaching the crisis level. NATCA also stated that the FAA missed their hiring goals for five consecutive years. The academy has been operating well below capacity in recent years due to the COVID pandemic and other issues.

As a result of the DCA accident, there is a concerted effort among Congress and the FAA to reach staffing goals. Transportation Secretary Sean Duffy recently announced cash incentives to "supercharge" the controller workforce. Academy graduates and new recruits who complete the initial qualification training will receive $5,000. Additionally, academy graduates who are assigned to one of the thirteen "hard-to-staff" facilities will receive $10,000. Shortly after these incentives were announced, a record setting government shutdown forced controllers to work without pay for more than a month creating a surge in early retirements.

In the next chapter I will address the factual details that led to the DCA collision, but before I do it would be helpful to understand the various levels of air traffic control. The airplanes that you see traversing the sky while leaving behind their tell-tale white contrails are flying at very high altitudes. They are under the control of an Air Route Traffic Control Center (ARTCC) commonly referred to as "the center." There are 23 of these control centers across the contiguous United States. One of these is the Washington Center and it covers nearly 160,000 square miles of airspace over parts of Pennsylvania, Maryland, Delaware, the District of Columbia, Virginia, West Virginia, and North Carolina.

Airplanes flying at lower altitudes are under the control of a Terminal Radar Approach Control (TRACON) facility. DCA falls under the control of Potomac TRACON. This facility covers more than 20,000 square miles and manages arrivals and departures at

Washington National, Washington Dulles, Baltimore-Washington International, and Richmond International, as well as many smaller airports in between. Potomac TRACON is the fourth largest approach control facility in the US and handles more than 2 million flights each year.

Airplanes that are operating within about 5 miles of the airport are controlled by the airport control tower. There is more than one controller in the tower at a busy airport like DCA. The local controller, who pilots refer to as the tower controller, is responsible for aircraft that are on the runways as well as airborne traffic that is in close proximity to the runways. There is a second local controller at DCA to monitor helicopter traffic, but that position was not staffed at the time of the midair collision. An extra set of eyes and ears are provided by a local assist controller who assists the local controller as needed. The ground controller is responsible for all traffic that is on the airport surface except for traffic on the runways. There is an operations supervisor or controller-in-charge that oversees all tower operations and there is a flight data/clearance delivery position that coordinates with other ATC facilities and issues flight plan clearances to pilots before each flight.

Here is an example of how traffic flows through the National Airspace System (NAS) to arrive at DCA. Washington Center receives traffic from the surrounding centers. Potomac Approach Control receives traffic from five different sectors within the Washington Center and maneuvers that traffic, through a series of turns and speed restrictions, into a single line to approach DCA. To accommodate the high number of arrivals, the airplanes are spaced 4 miles apart. This allows just enough room for one airplane to land and clear the runway, another airplane to take off, followed by the next arrival. About 5 miles from the runway, control of these airplanes is transferred from Potomac Approach to the DCA local controller.

There are several variables that can disrupt the close spacing of airplanes on final approach. When this happens, the local controller has few options. He can direct an arriving airplane to "go around" with instructions to make a climbing turn and contact Potomac Approach to get back in line to start the whole process over again. Another option is to have the approaching airplane make a series of "S" turns to create additional space, but this can only be done in Visual Meteorological Conditions (VMC). The last option, which is only available for north operations, is to ask the arriving flight to switch to runway 33. This is the option that ultimately led to the collision between PSA 5342 and PAT 25.

When conditions repeatedly disrupt final approach spacing, the flow of traffic must be reduced. The DCA operations supervisor will notify Potomac Approach that greater spacing is required. This, in turn, can disrupt the flow of traffic for Potomac Approach which, in extreme conditions, can lead to a "ground stop" meaning that flights that have not yet departed for DCA are held on the ground until conditions improve.

The hourly traffic rate at DCA is the only variable that can be controlled. There is no room to expand the airport and the prohibited airspace that surrounds the airport cannot be eliminated. DCA tower controllers and Potomac approach controllers have raised safety concerns about the high number of hourly flights, but those concerns have fallen on deaf ears as reported during the public hearing that is covered in the nest chapter.

Chapter 8
The NTSB Investigation

THE NATIONAL TRANSPORTATION SAFETY BOARD (NTSB) is an independent government agency tasked with promoting safety and investigating transportation accidents. Their investigations determine the probable cause of accidents from which they make recommendations to prevent similar accidents from occurring again. It is up to the FAA, and the aviation community, to act on their recommendations. The NTSB does not have regulatory authority, and they do not assign blame or liability, but their investigations lead to changes that make aviation safer for the traveling public. Since its inception in 1967, the NTSB has investigated thousands of aviation accidents and surface transportation events such as the oil pipeline rupture that dumped one million gallons of crude oil into the Kalamazoo River, the toxic train derailment in East Palestine, Ohio, and the tragic collapse of the Francis Scott Key Bridge in Baltimore, Maryland.

The NTSB considers every possible source of information when investigating an accident including, but not limited to, eye-witness accounts, witness interviews, video footage, ATC radar and voice recordings and the airplanes' recording devices, commonly referred to as the" black boxes." Investigations begin immediately following an accident and can take months or years to conclude. Investigators were on the scene of the DCA accident quickly since their headquarters are nearby. A group of experts ranging from pilots to aircraft mechanics, air traffic controllers to meteorologists, and human performance researchers to data specialists are involved in the investigation. They begin by collecting evidence and conducting interviews. This phase of the investigation of the DCA crash took about six months, which is typical.

The next phase of the investigation involved a public hearing in a formal setting. The DCA hearing included nearly thirty hours of testimony during a three-day period. The second day of the hearing lasted nearly twelve hours. This hearing is a fact-finding mission that gives representatives from the NTSB, and other interested parties, the opportunity to question groups of expert witnesses to better understand the complexities of the accident. Expert witnesses are selected based on their ability to provide the most accurate information. They testify under oath because their testimony becomes part of the permanent record. The NTSB is an evidence-based organization, so they consider only verifiable facts and not opinion or speculation. Several times during the public hearing NTSB Chair Jennifer Homendy interrupted a witness to remind them, "no speculation please."

Since the hearing was open to the public, some family members of loved ones who perished in this horrific accident were in attendance. With heartfelt empathy, Chair Homendy acknowledged their presence and express genuine sorrow for their loss. Families were given the opportunity to leave the room when sensitive information was discussed and the American Red Cross and counselors were on hand to assist these families throughout the hearing.

Portions of the hearing would be difficult to follow for someone with no aviation background, so I have translated the technical testimony into layman's terms to give you a better understanding. The entire hearing can be viewed on the NTSB YouTube channel. The hearing was divided into five categories. Each category consisted of a panel of technical experts who asked questions to a group of expert witnesses on the following topics:

Panel 1 – Overview of Accident Helicopter's Air Data Systems and Altimeters

Panel 2 – Overview of the DCA Class B airspace and Helicopter Routes

Panel 3 – Training, Guidance and Procedures Applicable to DCA Air Traffic Control

Panel 4 – Overview of Collision Avoidance Technology

Panel 5 – Safety Data and Safety Management Systems at the Various Organizations

Helicopter operations and Air Traffic Control procedures dominated much of the three-day hearing. The following is a summary of what I consider to be the most significant information that the hearing revealed.

A high volume of helicopter traffic has long complicated the airspace around DCA. FAA statistics revealed that during a two-year period beginning in 2017, over 50 helicopter operators conducted approximately 86,000 helicopter flights within 30 miles of DCA. Most of these flights did not originate or terminate at DCA but they shared the airspace with the thousands of passenger planes that did.

Helicopter routes were established nearly forty years ago to organize helicopter traffic into a predictable pattern, lowering the workload for controllers. For example, PAT 25 was cleared to follow Route 1 to Route 4, then proceed south, requiring a minimal amount of communication with the control tower but, the hearing revealed alarming and unsettling information about these routes.

The accident occurred on helicopter Route 4, which crossed the final approach path to runway 33. The investigation revealed that operations along helicopter routes are not well defined or understood by pilots, controllers, and even the FAA. Helicopter Route 4 specified an altitude of 200 feet, which most operators interpret as a maximum altitude. Mr. Dressler, a former Army helicopter pilot, a retired Air Force officer, and currently an official with Metro Aviation, a prominent medevac operator serving DC hospitals, testified that his pilots maintain 200 feet when navigating along Route 4. Chief Warrant Officer VanVechten, a former standardization instructor with the 12th Aviation Battalion, testified that the Army expects its pilots to fly the charted altitude with a tolerance of plus or minus 100 feet. Mr. Allen, the DCA control tower manager at the time of the accident, testified that controllers expect pilots to maintain 200 feet or below. Mr. Fuller, the Acting Chief Operating Officer of ATC Operations, testified that these altitudes are recommended but not regulatory.

The helicopter route chart legend gives conflicting information. One section of the legend refers to the altitudes as recommended while another states that they are mandatory. The chart legend is being updated to reflect the correct information which is, these altitudes represent mandatory maximum limits. Additionally, Letters of

Agreement between the helicopter operators that use these routes and the Air Traffic Control Tower state that these altitudes are maximum limits.

Route 4, as it was depicted, followed the east bank of the Potomac River adjacent to the airport. Mr. Dressler testified that his pilots "hug the shoreline" as depicted when navigating along Route 4. Mr. VanVechten testified that it was "tribal knowledge" for his pilots to follow the shoreline. Mr. Allen testified that controllers expect pilots to follow the shoreline of the river as depicted. The crew of PAT 25 was flying near the middle of the river when they ran into the CRJ but, Mr. Fuller testified that helicopter routes have no lateral limits meaning that following the shoreline is merely a recommendation. The FAA representative for aeronautical charting, Ms. Murphy, testified that the charted helicopter routes are nothing more than a symbol used to showcase the routes pictorially. Since these routes are not clearly defined by navigation aids, they have no lateral limits. They are depicted only on visual navigation charts to highlight the recommended flight path.

The following conversation was taken from the accident helicopter's cockpit voice recorder seconds before the collision. The Conflict Alert had just been triggered in the control tower prompting the controller to verify that the crew on PAT 25 still had the CRJ in sight and was maintaining visual separation. Chief Warrant Officer Eaves was the instructor, and Captain Lobach was the flying pilot on this recurrent proficiency flight.

Mr. Eaves – alright kinda come left for me ma'am I think that's why he's asking.

Captain Lobach – sure.

Mr. Eaves – we're kinda…

Captain Lobach – okay. fine.

Mr. Eaves – …out towards the middle.

Sadly, those were the last words spoken by the crew of PAT 25.

This led to a lengthy round of questioning that I have paraphrased and condensed below:

NTSB – How was a pilot expected to navigate Route 4?

FAA - In accordance with their air traffic control clearance.

NTSB – But their clearance was to follow the helicopter route. How are they separated from other traffic?

FAA – It is up to ATC to provide separation.

NTSB – Is a pilot expected to maintain the altitude that was depicted on the helicopter route?

FAA – If the altitude was assigned by ATC, then the pilot would be expected to adhere to that altitude.

NTSB – Mr. VanVechten, is it your understanding that helicopter route 4 provided vertical separation from the approach path to runway 33?

Mr. VanVechten – Yes, sir.

NTSB – Did you ever cross the final approach path when an airplane was landing on runway 33?

Mr. VanVechten – No sir. I was always instructed by ATC to hold until the airplane was clear and then allowed to proceed.

NTSB – Mr. Allen was that the procedure that was in use at the time of the accident?

Mr. Allen – No sir that was a technique used by some controllers, but it was not a procedure.

I was dumbfounded to learn that these helicopter routes, which had been in place for many years, were so ill-defined and ambiguous that even the FAA could not accurately describe their limitations. Helicopter routes were never intended to provide separation between helicopters and airplanes, but many pilots thought that they did. It is the responsibility of the controller to provide safe separation. The fact that some controllers would not allow a helicopter to proceed along Route 4 when an airplane was approaching runway 33 is testament to the inherent risk associated with Route 4. You might ask, why didn't these controllers voice their concerns? The alarming answer is, they did!

In 2013, twelve years prior to this tragic accident, a near miss between an Army helicopter that was transiting Route 4, and a passenger plane that was approaching runway 33, occurred at the very spot where PAT 25 ran into PSA 5342. Following this near tragedy, DCA tower controllers formed a helicopter working group to address safety concerns. They recommended that the FAA shut down, move, or modify Route 4. Their recommendation was rejected citing "Continuity of Government." Since then, the number of close calls has continued to climb and the helicopter working group has grown to include all the helicopter operators in the region.

Mr. Allen testified that prior to the accident, he had been part of the helicopter working group for several years and once again proposed relocating Route 4. The FAA cited reasons why that was not possible, so Mr. Allen then proposed adding hot spots to the chart to highlight high traffic areas. This proposal was also rejected because there is no definition for a hot spot on an aerial chart. Bear in mind that each of these proposals had to be submitted through formal channels and apparently never made it higher than the lowest levels of the FAA hierarchy. Mr. Allen then proposed adding caution notes to the chart. NTSB Board member Inman pointed out that the Los Angeles helicopter charts contain a caution note for intense helicopter operations. He asked Ms. Murphy if similar caution notes could be added to the DCA area charts and she replied yes. Mr. Inman asked, why haven't you done that and she replied, that's not what they asked for, they asked for hot spots. Out of frustration, Chair Homendy intervened and said you are telling us that your bureaucratic process will not allow you to make changes. She went on to chastise the FAA for their lack of action. "Every sign was there that there was a safety risk, and the tower was telling you that…so you transferred people out rather than taking ownership over the fact that everybody in the FAA in the tower was telling you that there was a problem. [We added up how many steps it takes to get from the tower to FAA headquarters to get a policy change, 21 steps.] Fix it. Do better."

Mr. Allen and three other management personnel were transferred after this accident which explains Chair Hominy's remark that "you transferred people out rather than taking ownership." In fact, the DCA Tower Manager has been replaced 10 times in the last 12 years. NTSB investigators learned that the helicopter routes were supposed to be reviewed every year, but they could find no documentation that those reviews had ever been accomplished, and they were unable to determine who was responsible for conducting those reviews.

The investigation also revealed that the Army Black Hawk pilots receive no training on airport or airline operations at DCA. This became apparent when Mr. VanVechten testified that he had heard controllers' clear airplanes for the Mount Vernon Visual Approach, but he could only guess what that procedure entailed. He also testified that he did not know that airplanes would be circling to runway 33. He assumed that they would be flying a straight-in approach. When asked if he was surprised to learn that a helicopter proceeding along the east bank of the river at 200 feet would cross within 75 feet of an approaching airplane, he replied, if I had known that I would have instructed my pilots to hold short whenever an airplane was approaching.

Dr. Casner, a research psychologist for NASA, testified that it is vitally important for pilots to know where to scan the sky in order to spot traffic. He explained that although peripheral vision allows us to see objects up to 180 degrees, we can only focus in a very narrow range of about 3 degrees. He cited a study that determined that pilots could spot traffic 8 times faster if they knew about where to look. When the ATC Conflict Alert was sounded for the second time, the local controller asked the helicopter crew if they still had the CRJ in sight. If the controller had said, do you still have the CRJ in sight at your 10 o'clock position, the helicopter pilots would have looked to their left and in all likelihood would have spotted the CRJ in time to avoid the collision.

At the time of the accident, the crew on PAT 25was conducting NVG (Night Vision Goggle) proficiency training which restricted their ability to scan for traffic. Mr. VanVechten was questioned about the use of NVGs and explained their advantages and disadvantages. The advantage of NVGs is the ability to see things in the dark that would otherwise be invisible. Disadvantages include decreased depth perception and a narrow field of view. Without goggles peripheral vision allows the average person to see about 90 degrees to either side when looking straight ahead. With goggles peripheral vision is reduced to about 20 degrees either side. This is like looking through a pair of binoculars which allow you to see straight ahead quite well but not to the side. The CRJ was approaching from left to right and would have been outside of the pilot's field of view when looking straight ahead. Mr. VanVechten stated that these goggles hang from the pilot's helmet and do not fit tight against the eyes. This allows pilots to look below the goggles to scan their flight instruments and to scan for traffic with the naked eye.

Additionally, NVGs are monochromatic so the red, green, and white lights that outline the silhouette of an airplane would appear as various shades of green. Mr. VanVechten also testified that it can be difficult to distinguish lights on the ground from airborne lights. He stated that flying with NVGs is more difficult around cultural (ground) lighting, like that which surrounds DCA. Chief Warrant Officer Eaves accepted responsibility for maintaining visual separation from the CRJ but failed to do so. There was more than one airplane approaching DCA via the Mount Vernon visual approach at the time of the accident and Mr. Eaves must have focused on the wrong airplane.

An Army standardization instructor pilot testified, post-accident, that he and other pilots he knew would request visual separation from traffic that they had not yet visually acquired. I can only speculate that these pilots assumed the helicopter routes provided safe separation and that they would visually identify the traffic when they got closer. Traffic advisories were issued so frequently at DCA that they may have led to complacency by some pilots and controllers who assumed that the mere fact that they were following the helicopter route would keep them safe. Visual separation is used every day throughout the National Airspace System, and it is a useful tool, but it has limitations.

Figure 8 represents an NTSB simulation of Chief Warrant Officer Eaves' view from the right seat of the helicopter. The right and center windshields, which are outlined by a thin white line, illustrate the field of view when not wearing NVGs. Note that PSA 5342 is clearly visible in the center windshield. The circle illustrates the field of view as seen through NVGs. Note that the CRJ is not in view when looking through the NVGs, but another airplane is visible in the distance near the left edge of the circle. Figure 9 illustrates Captain Lobach's view from the left seat.

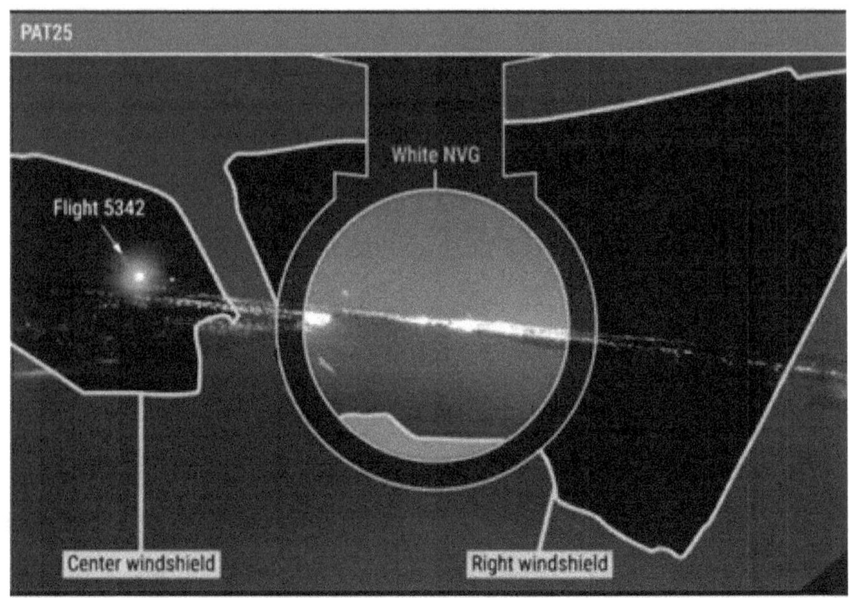

Figure 8. This is an NTSB simulation of Chief Warrant Officer Eaves' view from the right seat of the helicopter.

Source: NTSB Final Report

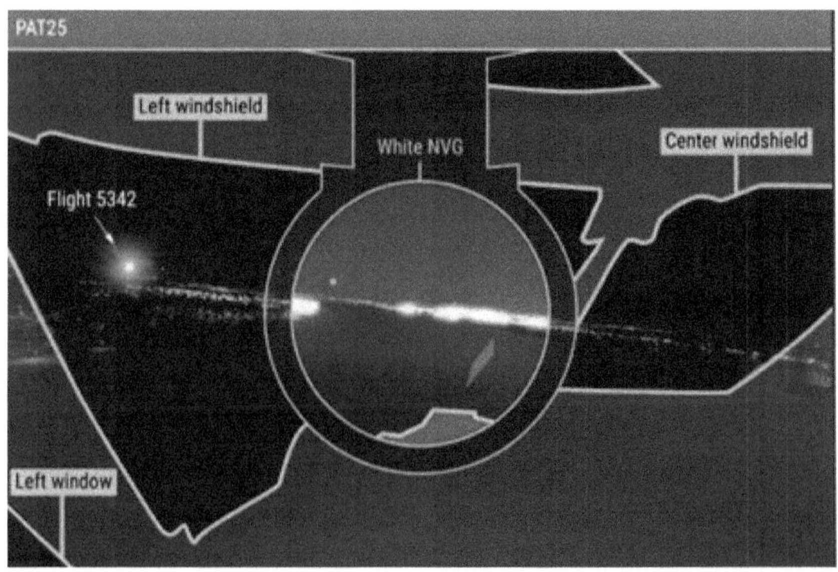

Figure 9. A simulation of Captain Lobach's view from the left seat of the helicopter.

Source: NTSB Final Report

Adding to the list of contributing factors, post-accident investigation revealed that the helicopter's altimeter was reading 80 to 100 feet low. In other words, when the altimeter read 200 feet, the actual altitude was between 280 and 300 feet. The collision occurred at 278 feet. Altitude is displayed to pilots in one of three ways. The primary instrument that is referenced for most flight operations is a barometric altimeter which senses changes in atmospheric pressure and converts that information into feet above sea level. This is the instrument that gave erroneous information to the helicopter pilots. Next is a radar altimeter which uses radio waves to determine aircraft height above ground level. Airplane pilots use this information to determine their height above the runway during low visibility instrument approaches. Helicopter pilots use this information during landing and hovering. It is not used during point-to-point navigation because ground elevation is constantly changing. The third indication is derived from the GPS. This is only used as a reference but can be helpful in non-normal situations.

All three sources of altitude information on the CRJ were determined to be within limits. The GPS altitude for the helicopter, which is derived from ADS-B, could not be determined because its ADS-B was not transmitting. The NTSB tested three additional Black Hawk helicopters from the 12th Combat Aviation Battalion and determined that their barometric altimeters indicated 80 to 130 feet low. It is difficult to accurately measure static barometric pressure on helicopters due to "rotor wash" which is the downward rush of air caused by the rotor blades. At cruising altitudes this error is somewhat negligible but less forgiving when flying at low altitude. The investigation revealed that Army pilots were not made aware of this anomaly.

When asked about pilot experience, Mr. VanVechten stated that five years ago the Army was attracting pilots with thousands of hours, but today's recruits have just hundreds of hours. By comparison, Mr. Dressler stated his company has a 2,000-hour minimum requirement for new medevac pilots. In earlier interviews, some pilots from the 12th Combat Aviation Battalion reported that it was difficult to reach the minimum flying requirement of 48 hours every six months. Mr. VanVechten stated that scheduling logistics, adverse weather, and other factors make it difficult for pilots to maintain the minimum flight hour requirements. Airline pilots typically fly about 85 hours each month. Of course, airlines schedule their pilots for maximum productivity whereas military pilots fly to maintain currency in case their services are needed. Nonetheless, there is a significant difference in the experience level of these two pilot groups.

The Black Hawk pilot, Captain Lobach, had logged about 500 hours of total flights time and five hours in the month prior to the accident. The hearing revealed that Captain Lobach had some issues with vertigo in the past. About one hour before the accident, Mr. Eaves took control of the helicopter for about ten minutes. During that time the Cockpit Voice Recorder captured Captain Lobach saying, "I don't know about you, but I'm feeling a little dizzy looking down and around with the winds or whatever it is but not too bad." Mr. VanVechten stated that experiencing vertigo when looking through Night Vision Goggles is not uncommon, especially for the pilot that is not physically flying the aircraft. This may not have been a factor in this accident, but it begs the question, why is training like this taking place near passenger planes.

The Black Hawk instructor, Chief Warrant Officer Eaves, had about 1,000 hours of total flying time. Mr. VanVechten stated that Warrant Officers focus only on flying while commissioned Officers have staff duties in addition to flying.

The Bravo Company of the 12th Combat Aviation Battalion, the unit involved in this accident, is equipped with "M" model and "L" model Black Hawk helicopters. The "M" models have more sophisticated instrumentation and an autopilot that can maintain altitude. Captain Lobach learned to fly in the "M" model but was subsequently assigned to the "L" model, which is arguably more challenging to fly. Mr. VanVechten testified that very few pilots fly enough hours to maintain currency on both models.

Moving on to air traffic control procedures, Mr. Allen stated that the practice of transitioning airplanes from runway 1 to runway 33, to gain space between arriving and departing aircraft, is a common technique. Controllers call this "offloading to 33." Mr. Allen also testified that controllers use every tool at their disposal to handle the high volume of hourly traffic that plagues DCA every day. In his words, we just "make it work." In addition to offloading to 33, controllers use instructions, such as slow to your minimum speed, no delay exiting the runway, and be ready for an immediate takeoff, to keep traffic moving. When asked if these tactics were jeopardizing safety Mr. Allen replied, "they're pushing the limits." He also stated that "it can be taxing on a person, you know, constantly having to give, give, give, or push, push, push in order to efficiently move traffic.

Being a high volume, high complex airport with not a lot of real estate, you have to keep things moving."

DCA is a hub for American Airlines. A hub refers to the "hub and spoke" model that airlines use to generate multiple connecting flights. Imagine, for example, a wheel with fifteen spokes and each spoke represents a different city. The hub, or center of the wheel represents DCA. Flights depart each of these cities and arrive at DCA within minutes of each other. In less than one hour they all depart for different cities thus creating two hundred and twenty-five combinations of connecting flights. This is great for travelers but bad for traffic congestion because these flights arrive in "waves" rather than being evenly distributed throughout the hour.

As we have already discussed, DCA utilizes a "slot" system to control congestion. This system is based on clock hours, for example, thirty landings might be allowed between 9 a.m. and 10 a.m., and another thirty between 10 a.m. and 11a.m., and so on. It would be ideal if those flights were evenly distributed throughout the hour but that is not the case. Mr. Lehman, a manager at Potomac TRACON, the approach control facility that feeds traffic to the DCA tower, testified that American Airlines uses a strategy called "front loading and back loading" of their schedule. For example, if they are authorized fifteen landings per hour, ideally those flights would be spaced evenly throughout that hour. Front and backloading refers to scheduling as many arrivals as possible at the end of one hour, and again at the beginning of the next. This results in more than fifteen landings in a sixty-minute period but within the limit of fifteen landings during each clock hour. This strategy creates more connecting opportunities for passengers but also creates a surge in traffic that forces pilots and controllers to "make it work." New York's LaGuardia airport experienced this same issue, so they changed their slot allocations to 30-minute intervals to lessen the effect of front and back loading.

In May of 2023, Mr. Lehman proposed lowering the arrival rate at DCA. Based on an FAA supplied table he proposed lowering the arrival rate for north operations from thirty-six to thirty-two arrivals per hour. Ten days later he was advised by Ms. Chendi, the Traffic Management Officer for the Washington District, that we are not going to move in that direction. She mentioned that this is a bad time to bring this up because of the pending Reauthorization Act. That is the bill that added five additional round trips to DCA just eight months prior to the accident. She also stated that her office determined that runway 33 is being underused so she reached out to the airlines to verify that their pilots could land on this runway. She described runway 33 as "capacity that is not being used" implying that more frequent use of this runway would facilitate the high number of hourly flights. She also stated that pilots" just choose to use runway 1."

With all due respect to Ms. Chendi, it is my opinion that pilots "just choose to use runway 1" because runway 33 is a poor choice due to the lack of approach guidance and its relatively short landing length. However, it is an option for smaller airplanes or when a strong north-west wind is blowing such as on the night of this accident. For most situations, the best option for passenger plane pilots is to follow the precision approach guidance to runway 1 and touch down with the equivalent of six football fields of additional stopping distance. The FAA is quick to point out that pilots have the option to not accept runway 33 but I submit that pilots and controllers should not be put in a position where they must use improvised procedures to “make it work.” Runway 33 should not be used as a tool to mitigate airport overcrowding.

Later during the hearing, Ms. Chendi clarified her position on arrival rates. She stated, if runway 1 was the only available runway, she would be in favor of reducing the arrival rate but, since runway 33 is an available option, it should be used. I completely disagree. Runway 33 is an available option under very limited conditions for “some” airplanes. That is hardly a solution for the hourly surge of traffic that plagues DCA every day. Board member Inman asked Mr. Lehman, how can we fix this? Mr. Lehman replied, listen to our concerns. We are bringing legitimate safety concerns to FAA management, and we are not being heard.

Several times throughout the investigation, the topic of transparency at the FAA was raised. On some occasions the NTSB had to submit numerous requests for documents. Chair Homendy stated that she had to personally reach out to the FAA Administrator and the Secretary of Transportation and finally received thousands of documents a few days prior to the public hearing. At one point during the hearing when Ms. Chendi was testifying about the arrival rate at DCA, Mr. Fuller clearly “elbowed” her, and she stopped talking. This was brought to the attention of Chair Homendy, and she stated that she is not going to draw any assumptions, but she is not going to put up with that. When the hearing resumed after a short break, Mr. Fuller and Ms. Chendi were seated at opposite ends of the witness’ table. Chair Homendy also stated that some FAA employees were reluctant to testify due to fear of retribution.

Individual interviews conducted by the NTSB prior to the public hearing revealed that some pilots accept the transition to runway 33 because they feel that they are helping ATC. PSA, in collaboration with the NTSB, and the FAA, conducted post-accident simulations to rank the difficulty of transitioning from runway 1 to runway 33. A PSA Captain and First Officer were selected to fly the simulations because they had similar flight hours and experience flying to DCA as Captain Campos and First Officer Lilley. When flying the Mount Vernon Visual Approach and landing on runway 1, both pilots ranked the workload as 1, using NASA's Bedford Workload Scale. A rating of 1 is defines as "workload insignificant." When flying the same approach with a transition to runway 33, The Captain ranked the workload as a 5 which is defined as "reduced spare capacity, and additional tasks cannot be given the desired amount of attention." The First Officer ranked this as a 7 which is "very little spare capacity, but the maintenance of effort in the primary task is in question." Finally, when the transition to runway 33 was flown with a Traffic Alert, as was the case for PSA 5342, both pilots ranked the workload a 7. There is good reason why runway 33 accounts for just four percent of the landings when DCA is in a north operation.

Due to the high volume of helicopter traffic near DCA, a second local controller position was established years ago to focus solely on helicopter traffic. This position is staffed Monday through Friday from 10 a.m. until 9:30 p.m. Staffing this position was mandatory prior to 2023 but since then the two controller positions can be combined under certain conditions, such as low traffic volume, adverse weather that keeps helicopters grounded, and staffing levels. I understand the first two reasons but, staffing levels? It has been widely reported, especially since this accident, that the ATC system is significantly understaffed. The helicopter position was staffed for only one hour and twenty minutes on the day of the accident but not at the time of the accident. Mr. Allen estimated that this position is not staffed more than fifty percent of the time.

Even though the helicopter crew reported that the CRJ was in sight, and that they would maintain visual separation, the lone local controller failed to make a required and critical transmission. He failed to notify the CRJ crew about the helicopter. Per air traffic rules, "If aircraft are on converging courses, inform the other aircraft of the traffic and that visual separation is being applied." Without that information, Captain Campos and First Officer Liley would have no reason to suspect a traffic conflict so close to touchdown. They would be focused on the airplane that was taking off on runway 1 and about to cross in front of them.

The DCA local controllers communicate with helicopter traffic and airplane traffic on separate frequencies to reduce radio congestion. This works well when both local control positions are staffed, but when the positions are combined, the lone controller continues to use separate frequencies. He can transmit and receive on both frequencies simultaneously, but airplane pilots cannot hear responses from helicopter pilots and helicopter pilots cannot hear responses from airplane pilots. This is somewhat like listening to one side of a phone call while not hearing the other and it can lead to a reduction in situational awareness.

Radio communications at a busy airport like DCA can be chaotic. Sometimes transmissions are blocked by two people transmitting at the same time. After the Conflict Alert was triggered and the local controller verified that PAT 25 still had the CRJ in sight, he issued the following clearance: "PAT 25 pass behind the CRJ." Most of that transmission was blocked so the helicopter crew heard "PAT 25………CRJ." The accident occurred a few seconds later.

In my opinion, from the controller's point of view, the helicopter looked as though it might pass in front of the CRJ, since it was not following the shoreline. This explains why the controller said, "pass behind the CRJ." To pass behind the CRJ, PAT 25 would have to slow down or begin a left turn, but no action would be necessary if the helicopter pilots thought that the CRJ was at their 12 o'clock position. This is the only conclusion that connects all of the elements leading up to the collision.

The investigation revealed that the local controller, who had been qualified in that position for only six months, felt overwhelmed about 10 to 15 minutes prior to the collision. The operations supervisor, who was an experienced controller but only recently qualified as a supervisor, felt that the local controller did not need assistance. The investigation revealed that there were enough controllers on duty at the time of the accident to staff the helicopter position, but the supervisor did not think that it was necessary. The assistant local controller, who should have provided a second set of eyes and ears to assist the local controller, was performing nonessential duties at the time of the accident.

During the moments leading up to the collision, the Potomac Approach Control manager contacted the DCA operations supervisor to request decreased spacing and the supervisor approved his request. To be clear, decreased spacing means an increase in the number of arrivals. Just prior to the accident, the lone local controller was working with six airplanes and five helicopters.

When the Conflict Alert was triggered, the local controller verified that PAT 25 still had the CRJ in sight. He then turned his attention to the next arriving airplane. Two aircraft can safely pass inside of the parameters that trigger this alert provided one pilot clearly sees and avoids the other. The automated alert has no way to know that visual separation is being applied, and it cannot be silenced, so it becomes a "nuisance" alert. Nuisance alerts can lead to "alarm fatigue" which is a well-known phenomenon in healthcare where the constant beeping of monitors can result in delayed action, or sometimes no action at all. The investigation revealed that five conflict Alerts were triggered during an eighteen-minute period leading to the collision.

Investigators are concerned that the tower controllers had an overreliance on visual separation to mitigate the high number of conflict alerts that plague DCA every day. We already know, from the tragic 1978 midair collision over San Diego, that visual separation is not foolproof. Improved aircraft tracking technology is available today, but it will take years to implement. In the interim, controllers should take positive action and not rely on pilot provided visual separation alone.

Air traffic control facilities are ranked according to their complexity on a scale of 4 through 12 with 12 being the most complex. Controller pay is proportionate to the level of complexity. It makes sense that a controller at a busy, complex facility would make a higher salary than a controller at a small facility with a low traffic count. In 2018, the DCA tower was downgraded from Level 10 to Level 9 and Potomac Approach was upgraded from Level 10 to Level 12. This was the result of shifting the helicopter count from the DCA tower to Potomac Approach. As a result of this downgrade, DCA has found it difficult to attract experienced controllers, new controllers transfer out once they are fully qualified, and morale has decreased for those that remain. In my opinion, both facilities should be Level 12.

Post accident interviews revealed that the DCA tower controllers were not familiar with Threat Error Management (TEM) training. TEM training was developed in 1994 by the University of Texas Human Factors Research Project in conjunction with Delta Airlines in response to accidents that are caused by human factors. TEM teaches pilots to assess and identify potential threats and to take proactive, rather than reactive, steps to mitigate those threats. It has played an integral part in pilot training for more than twenty-five years. More recently, it was integrated into air traffic controller training. Mr. Fuller believes that controllers received this training but might not recognize it by name.

Data was available, prior to this accident, that provided evidence of midair collision risk between airplanes and helicopters at DCA but access to that data was limited.

The Army's safety reporting systems did not provide the organization with information about close encounters between Army helicopters and other aircraft that were later found to have occurred frequently.

In the next chapter I will review the results of the NTSB investigation which details their conclusions, the probable cause of the accident, and their recommendations to prevent future accidents.

Chapter 9
The NTSB Final Report

THE NTSB FINAL REPORT is the culmination of twelve months of exhaustive investigation to determine the probable cause of the accident and to make recommendations to prevent a similar accident from occurring again. The report follows a standard format beginning with a summary of the accident. The main body of the report details every aspect of the investigation such as pilot and controller qualifications, regulatory compliance, aircraft maintenance records, meteorological conditions and more. The final pages of the report reveal the conclusions, probable cause, and recommendations made by the NTSB. In this chapter I will highlight the most significant findings, but the entire report is public information and can be accessed on the internet by searching NTSB report 26-02.

Conclusions

The final report details nearly 400 pages of research that led to the conclusions that I have condensed and paraphrased below. The entire text of these conclusions is appended.

The crew members on both aircraft and the DCA tower controllers were properly certificated and qualified to perform their duties. Available evidence does not indicate that fatigue, medical conditions, or substance abuse were factors. However, the FAA's Air Traffic Organizations decision to not conduct drug testing as soon as possible, and to not conduct alcohol testing at all, violated DOT requirements.

Both aircraft were properly certificated, equipped, and maintained. Review of surveillance videos indicated that the exterior lighting on both aircraft was operating at the time of the collision. There was no evidence of any local atmospheric pressure anomalies that would have impacted barometric altimeter readings.

The control tower was adequately staffed at the time of the accident to cover both the helicopter and airplane local control positions had the operations supervisor chosen to do so. These positions should have been separated at the time of the accident given traffic volume and complexity. In the two minutes before the accident when traffic volume was increasing, the assistant local controller should have prioritized surveillance of aircraft to assist the local controller, rather than diverting her attention to administrative tasks, which could have been completed when traffic volume and complexity had subsided.

Due to extended time on position at the time of the collision and his complacency, the operations supervisor was likely experiencing reduced alertness and vigilance. The lack of mandatory relief periods for supervisory air traffic control personnel is contrary to human factors research that shows clear performance deterioration in situations of prolonged time on task.

Although the local controller provided an initial traffic advisory to the helicopter crew, he did not provide a corresponding advisory to the CRJ crew. If he had provided both aircraft with relative position and positive control instructions, the crew of either aircraft could have taken immediate action to avert the impending collision.

Scenario-based training in Threat and Error Management (TEM) would help controllers identify and mitigate risks and strengthen situational awareness. A risk assessment or decision-making tool would likely have benefited the accident operations supervisor in identifying and mitigating the operational risk factors that were present on the night of the accident.

Due to degraded radio reception, the helicopter crew did not receive salient information that the CRJ was circling to runway 33. The helicopter instructor pilot did not positively identify the CRJ at the time of the initial traffic advisory despite his statement that he had the traffic in sight and requested visual separation. With several other targets located directly in front of the helicopter represented by points of light with no other features by which to identify aircraft type, and without additional position information from the controller, the instructor pilot likely identified the wrong target. Radio interference, ambiguous visual cues, and the lack of TCAS or ADS-B In traffic alerts, likely reinforced the helicopter pilot's expectation bias that the CRJ was among the traffic approaching runway 1 and did not pose a conflict.

The lack of training on airplane operations and the mixed-traffic operating environment at DCA represented a safety vulnerability for Army flight crews.

It is likely that the helicopter altimeter read 100 feet low resulting in the crew erroneously believing that they were under the published maximum altitude for Route 4. The FAA and the Army failed to identify the incompatibility between low helicopter routes and the error tolerances of barometric altimeters, which contributed to helicopters regularly flying higher than published maximum altitudes.

The Army's post-installation functional check of the transponder on the accident helicopter was insufficient to detect that it was not broadcasting ADS-B Out. It is possible that incorrect settings may be present on other aircraft used throughout the Department of War armed services.

The CRJ crew did not see the helicopter until it was too late because of the high workload imposed during the final phase of their approach. The practice of "offloading" arrival traffic to runway 33 was a routine mitigation strategy to generate spacing that was not provided by Potomac Approach Control. Time-based flow management, or metering, which is now being used at DCA, but not at the time of the accident, would have provided a consistent flow of traffic with more accurate spacing and greater predictability, thereby reducing controller workload.

The DCA air traffic control tower has significant airspace, airfield, mixed fleet, and operations complexities that appear to be inconsistent with its current facility level classification

The FAA Air Traffic Organization (ATO) failed to recognize external compliance verification results as indicators of systemic traffic management, volume, and flow issues for which controllers were required to compensate.

The longstanding practice of relying on pilot-applied visual separation as the principal means of separating helicopter and fixed-wing traffic introduced unacceptable risk.

The practice of maintaining separate radio frequencies when the local and helicopter control positions were combined decreased overall situational awareness for pilots.

Providing controllers with additional salient cues regarding the perceived severity of a potential conflict would reduce controller cognitive load and would likely improve reaction time to the most critical conflict alerts.

Annual reviews of helicopter route charts, as required by the FAA, would have provided an opportunity to identify the risk posed by Route 4 but there is no evidence that these reviews were being carried out.

The information published by the FAA regarding helicopter routes was insufficient to provide operators with a complete understanding of the helicopter route structure and its lack of procedural separation from fixed-wing traffic. Current aeronautical charting does not provide information on VFR helicopter routes that may conflict with approach and departure corridors, reducing pilot situational awareness.

The lack of ADS-B Out from the accident helicopter did not contribute to this accident because the CRJ was not equipped with ADS-B In, which is not required.

The Army's standard operating procedures that prevent flight crews from enabling ADS-B Out, while not performing sensitive missions, limit their visibility by aircraft that are equipped with ADS-B In.

Although the airplane's TCAS operated as designed, it was ineffective in preventing the collision because the resolution advisory is inhibited at low altitude. The traffic advisory function is not inhibited, but if it included aural alerts, such as, TRAFFIC – 10 O'clock HIGH- 2 MILES, it could reduce the time pilots need to visually acquire target aircraft.

Had the CRJ been equipped with ADS-B In information, that showed directional traffic symbols, the crew would have received enhanced information about the risk posed by the helicopter, which could have enabled them to take earlier action to avert the collision.

Although the helicopter pilots were equipped with tablets that had the ability to display surrounding traffic, it is unlikely that the pilots were using them at the time of the accident due to the workload associated with low-altitude flight.

The next generation of TCAS is called ACAS or Airborne Collision Avoidance System. There are two variants of ACAS, Xa for aircraft and Xr for rotorcraft. ACAS Xa provides a lower inhibit altitude, an expanded alerting envelope, and reduced nuisance alerts over TCAS. Although not yet commercially available, had the helicopter been equipped with ACAS Xr with integrated aural alerting, the helicopter crew could have received an alert regarding the CRJ and could have taken action to avert the collision.

Prior to this accident, multiple data sources provided evidence of midair collision risk between airplanes and helicopters at DCA. However, limited access to this data hindered industry and government stakeholders' ability to identify and mitigate these risks. Improving access to information about aircraft close encounters would increase the likelihood of detecting and mitigating hazards before accidents occur. The FAA's lack of an established process to inform parties about their involvement in events such as near-midair collisions reduces the likelihood of fully understanding and mitigating future midair collision risk.

The FAA's Air Traffic Organization was made aware of and had multiple opportunities to identify the risk of a midair collision between airplanes and helicopters at DCA. However, they did not effectively communicate this information with external stakeholders.

Changes to standard operating procedures at DCA that removes the requirement for the operations supervisor (OS) to document the time and reason for combining or de-combining the helicopter local control position made it less likely that the OS would consider and evaluate the associated risks. Safety risk management practices were not fully integrated into DCA's operations and did not identify or mitigate the operational challenges faced by controllers or the lack of guidance regarding operational risk assessments for controllers and supervisors. Air Traffic Organization management did not support its workforce, encourage open communication, identify and mitigate risks, or foster a just culture, which eroded the overall safety culture within the ATO.

The Army did not have a flight safety data monitoring program for helicopters, and as a result, was unaware of routine altitude exceedances and related risks. Their safety reporting systems for pilots were not well utilized and did not provide the organization with information about close encounters between Army helicopters and other aircraft that were later found to have occurred frequently. Their process for allocating resources to aviation safety management did not ensure the development of a robust safety management system for helicopter operations. Their aviation safety system failed to consistently detect, interpret, and act on signals of latent hazards, resulting in degraded safety assurance, organizational learning, and safety culture.

Probable Cause

The above findings led investigators to conclude the probable cause of this accident which I have condensed and paraphrased below. The entire text is appended.

- The FAA's placement of a helicopter route in close proximity to a runway approach path, their failure to regularly review and evaluate helicopter routes and their failure to act on recommendations to mitigate the risk of a midair collision near DCA.
- The air traffic system's overreliance on visual separation without consideration for the limitations of the see-and-avoid concept.
- The lack of effective pilot-applied visual separation by the helicopter crew.
- Degraded performance by the tower controllers due to the high workload caused by combining the helicopter and local control positions, the misprioritization of duties among the controllers, and the lack of safety alerts to both flight crews.
- The Army's failure to ensure pilots were aware of the effects of error tolerances on barometric altimeters in their helicopters.

- The limitations of the traffic awareness and collision alerting systems on both aircraft, which precluded effective alerting of the impending collision to the flight crews at low altitude.
- An unsustainable airport arrival rate and airline scheduling practices, which regularly strained the DCA air traffic control tower workforce and degraded safety.
- The Army's lack of a fully implemented safety management system, which should have identified and addressed hazards associated with altitude exceedances on the helicopter routes.
- The FAA's failure to implement previous NTSB recommendations, including mandating ADS-B In for aircraft operating in congested airspace, and failure to follow its safety management system, which should have led to changes based on previously identified risks that were known to management.
- The absence of effective data sharing and analysis among the FAA, aircraft operators, and other relevant organizations.

Recommendations

As a result of this investigation, the National Transportation Safety Board made the following safety recommendations. Once again, I have condensed and paraphrased this information, but the full text is appended.

To the Federal Aviation Administration:

- Develop and implement time-on-position limitations for supervisory air traffic control personnel.
- Develop scenario-based training that trains controllers to continuously monitor their environment to more quickly and accurately identify threats.
- Develop and implement a risk assessment tool for supervisors.

- Initiate rulemaking that prescribes air carrier operation limitations at DCA in 30-minute periods, like those imposed at LaGuardia Airport.
- Implement operational use of the time-based flow management system.
- Reassess the airport arrival rate at DCA.
- Require each Class B or Class C air traffic control tower facility to evaluate its existing miles-in-trail procedures and make the results publicly available.
- Define objective criteria for the determination of air traffic facility levels to include cost of living.
- Determine whether the classification of the DCA air traffic control tower as a level 9 facility appropriately reflects the complexity of its operations.
- Develop a comprehensive training course on the proper use of visual separation.
- Conduct a comprehensive evaluation to determine the overall safety benefits and risks to requiring all aircraft to use the same frequency when the local control positions are combined.
- Implement technology that will alert controllers and flight crews when simultaneous transmissions are blocked.
- Develop and implement improvements to the conflict alert system to provide more salient and meaningful alerts to controllers based on the severity of the conflict triggering the alert and provide training to controllers on its use.
- Revise the Air Traffic Organization's initial event response procedures so that an appropriate on-site supervisor makes each postaccident and post incident drug and alcohol testing determination.
- Ensure that annual reviews of helicopter route charts are conducted throughout the National Airspace System.

- Conduct a safety risk management process to evaluate whether modifications to the remaining helicopter route structure in the vicinity of DCA are necessary.
- Amend helicopter route design to provide vertical separation from airport approach and departure paths.
- Review all existing helicopter routes to ensure alignment with updated criteria.
- Incorporate the lateral location and altitudes of helicopter routes on all instrument approach, visual approach, and departure procedure charts.
- Modify airborne collision avoidance system traffic advisory aural alerts to include clock position, relative altitude, range, and vertical tendency.
- Require new and existing traffic alerting and collision avoidance systems to integrate directional traffic symbols.
- Require all aircraft operating in airspace where ADS-B Out is required to also be equipped with ADS-B In.
- Require the use of ACAS X on new production aircraft that are subject to TCAS equipment regulations.
- Require existing aircraft that are subject to TCAS equipment regulations to be retrofitted with the appropriate variant of ACAS X.
- Evaluate the feasibility of decreasing the traffic advisory and resolution advisory inhibit altitudes in ACAS Xa systems to enable improved alerting throughout more of the flight envelope.
- If the inhibit altitudes can be safely decreased, require retrofitting of the applicable ACAS X variant on all aircraft that are subject to TCAS equipment regulations.
- Require that all rotorcraft operating in Class B airspace be equipped with ACAS Xr technology once the standard has been published.

- Create a public database of near-miss encounters that can be used to monitor their prevalence and identify areas of potential traffic conflict.
- Develop and implement a process that will notify involved parties after events such as near midair collisions or resolution advisory activations.
- Ensure that all safety management system functions and data sharing activities are conducted in collaboration with all relevant stakeholders.
- Establish a requirement that the operations supervisor or controller in-charge document when control positions are combined, along with a rationale for doing so.

To the US Army:

- Revise training procedures for flight crews to ensure that they receive initial and recurrent training on fixed-wing operations at DCA, including approach and departure paths, runway configurations, and the interaction of those traffic flows with published helicopter routes.
- Develop and implement a recurring procedure to verify the continued accuracy of recorded flight data.
- Incorporate information within the appropriate operator's manual for all applicable aircraft on the potential total error allowed by design that could occur in flight on an otherwise airworthy barometric altimeter.
- Develop and implement a transponder inspection procedure on all aircraft that ensures 1) the transponder ADS-B settings are correct, 2) the transponder is transmitting ADS-B, and 3) the transponder is transmitting the correctly assigned address.
- Establish a flight data monitoring program for Army helicopters
- Identify barriers to the utilization of flight safety reporting systems, develop a plan to address the identified barriers, and implement that plan across Army aviation units.

- Revise the method for allocating resources to ensure the development of a robust safety management system that will, at a minimum, identify and monitor the potential for midair collisions between Army aircraft and civil air traffic operating in the National Airspace System.
- Develop and maintain a flight safety management capability that ensures that this capability is both culturally and functionally integrated with units conducting sustained flight operations in the National Airspace System.

To the Department of War Policy Board on Federal Aviation:

- Conduct a study to evaluate the quality of radio transmissions and reception that adversely affects the safety of civilian and military flight operations.
- Require armed services to amend their operational procedures to allow flight crews to enable ADS-B Out while in flight.
- Require all military aircraft operating in the National Airspace System to be equipped with ADS-B In, including a cockpit display of traffic information that provides alerting audible to the pilot and that such requirement apply wherever ADS-B Out is required.

To the Department of Transportation:

- Require the FAA to demonstrate that each air traffic control facility it operates has the capability to accomplish required postaccident testing within two hours for alcohol and four hours for drugs.
- Work with the FAA Administrator to convene an independent panel to conduct a comprehensive review of the safety culture within the FAA's Air Traffic Organization and use the findings to enhance the ATO's existing safety management system and integrate it into all levels of the organization.

To the Department of Transportation Office of Inspector General:

- Complete an audit of the FAA Air Traffic Organization's safety management system functions and data sharing activities at all air traffic control facilities and determine whether these activities are conducted in collaboration with all relevant external stakeholders, ensuring that the audit's results are documented, reported to the Secretary of Transportation and the FAA Administrator, and made available to the public.

To the RTCA Program Management Committee:

- Finalize and publish the minimum operational performance standards for airborne collision avoidance system Xr for rotorcraft.

The accident investigation is now complete. As always, the NTSB left no stone unturned in determining the causes of the accident which in turn led to their recommendations. You could think of these recommendations as the NTSB's "wish list" for aviation safety. If they had regulatory authority, they could implement those recommendations and make aviation safer for all of us, but it is not that simple. It is up to Congress, and the stakeholders (FAA, Air Traffic Organization, airlines, Army, etc.) to respond to the recommendations and this is where, all too often, safety takes a back seat to economics.

Relatively inexpensive recommendations such as sharing safety data, logging time on position for tower supervisors, and training Army helicopter pilots about inherent altimeter errors will, in my opinion, be implemented quickly, but recommendations, such as mandating ADS-B In, and reducing arrival rates at DCA, will face an uphill battle because these changes will ultimately cost stakeholders millions of dollars. The recommendation to mandate ADS-B In for all aircraft has already stalled, as you will learn in the final chapter.

Chapter 10

Conclusion

THIS WAS A TERRIBLY HEARTBREAKING ACCIDENT. Nothing can be said or done to ease the pain and suffering felt by so many. Adding to that pain is the fact that this accident was so avoidable. It was not caused by violent weather or a catastrophic failure but by an industry that prioritizes profit over safety, a government bureaucracy that is stifled by its own regulations and a Congress that is too polarized to compromise.

The helicopter pilots clearly made a mistake by focusing on the wrong traffic, but it would be unfair not to consider the underlying factors that enabled their mistake. Chief Warrant Officer Eaves, the Pilot-In-Command/ Instructor on PAT 25, undeniably had his hands full. Expecting a pilot to conduct training in a complex helicopter flying 200 feet above the ground, adjacent to the nation's busiest runway, surrounded by passenger planes and peripherally blinded by night vision goggles is incomprehensible. Captain Lobach and Chief Warrant Officer Eaves made a deadly mistake, but a greater mistake was made by the Army for allowing training like this to take place in civilian airspace. Military pilots knowingly accept a certain degree of risk because of the nature of their mission, but airline passengers have a reasonable expectation to be transported safely without exposure to undue risk.

Military flight training, as we learned more than sixty years ago, should be conducted in a training area, not near a civilian airport. During a round of questioning at the public hearing, Brigadier General Brayman, the head of Army aviation at the time of the accident, asked if airlines conduct training at DCA. That struck me as an attempt by the General to justify the training mission of PAT 25 as somehow acceptable. Airline pilots spend many hours training, but we don't do it at a busy airport while surrounded by plane loads of passengers who place their trust in us to transport them safely. We train in simulators. Pilots are observed annually by a qualified instructor pilot during scheduled passenger flights, but this is observation only, no training takes place.

Captain Lobach and Chief Warrant Officer Eaves did not follow the shoreline as depicted by Route 4, but according to the FAA, that's okay because these routes are merely suggestions. Controllers and pilots testified during the public hearing that it was their understanding that the routes should be flown as depicted and for years, controllers have cleared pilots to proceed via these routes. How can there be such a disconnect between the FAA, who designed and charted these routes, and the pilots and controllers who rely on them?

There are two types of navigation charts. One for navigating under Instrument Flight Rules (IFR) and one for navigating under Visual Flight Rules (VFR). IFR charts depict routes, called airways, that must be flown exactly as depicted. VFR charts depict landmarks such as cities, bodies of water, and highways, to aid with visual navigation.

VFR pilots must comply with certain rules and regulations but for the most part they are free to navigate as they please. Helicopter routes are depicted only on VFR charts so technically they have no defined lateral limits, but this technicality is not well defined and is open to interpretation.

Airline flights are always flown under Instrument Flight Rules, even when the weather is clear, so airline pilots do not carry VFR charts, and they receive no training on helicopter operations around DCA. Military pilots receive no training on airline operations even though they share the same airspace with passenger planes every day. This is eerily similar to the 1958 midair collision over Las Vegas when civilian and military aircraft shared the same airspace without coordinating their operations.

Helicopter Route 4 provided reasonable separation from airplanes landing straight ahead on runway1, assuming that the helicopters remained close to the depicted path, but not for airplanes circling to runway 33. There is no defined procedure for flying the Mount Vernon Visual Approach and landing on runway 33. This is a "seat of the pants" maneuver that has been normalized to compensate for airport overcrowding. Even if Captain Lobach and Chief Warrant Officer Eaves were flying above the east bank of the river at 200 feet, and with an accurate altimeter, they would have no more than seventy-five feet of separation from the CRJ, which is completely intolerable, yet the FAA had been repeatedly warned of this hazard and took no action. Personally, I find it hard to believe that the working level employees at the FAA were in favor of ignoring those safety concerns. Those decisions were, in my opinion, made by upper management and were likely politically driven.

As a result of this horrific accident, some changes are finally taking place. Route 4 adjacent to the airport is permanently closed, controllers can no longer rely on visual separation near the airport, and military aircraft must broadcast ADS-B OUT data during non-critical phases of flight. Sadly, it takes an accident like this to effect change. Many safety enhancements are costly, making them difficult to mandate. As an example, changes were made in the aftermath of the 9/11 terrorist attacks. One such change involved reinforcing the cockpit door to prevent an intrusion. This was a significant upgrade but with a well-known weakness. There are times when that door must be opened in flight and flight crews have been forced to improvise safeguards to ensure the integrity of the cockpit.

No country takes airline security more seriously than the State of Israel. Their flag carrier, El Al Airlines, installed reinforced cockpit doors and secondary cockpit barriers on their airplanes long before the 9/11 attack. Thanks to the tireless work of Ellen Saracini, secondary barriers are finally becoming a reality for US airliners. Ellen is the widow of Captain Victor Saracini, the pilot of United 175, which was hijacked and crashed into the south tower of the World Trade Center after radical terrorists stormed the cockpit, murdered the pilots and took control of the airplane. The Saracini Aviation Safety Act was signed into law in 2018, but installation of the new barriers will not begin until 2026 and applies only to newly manufactured airplanes. Additionally, the Saraceni "Enhanced" Aviation Safety Act was signed into law in 2024 and will eventually retrofit secondary barriers on existing airplanes. This is long overdue, and sadly, typical of the uphill battle that safety advocates face all too often. To its credit, Southwest

Airlines is proactively installing these barriers on their new aircraft before the deadline.

If the helicopter had been transmitting ADS-B data, and the CRJ had been receiving that data, Captain Campos and First Officer Lilley would have had a 59 second warning before the collision. ADS-B Out has been required since 2020, notwithstanding the military exemption, but ADS-B In is not, despite repeated recommendations by the NTSB. A Senate proposed bill, the ROTOR Act (Rotorcraft Operations Transparency and Oversight Reform Act) closes loopholes that allow the military to operate certain missions without broadcasting their position and requires aircraft to be equipped with ADS-B In, but not until 2031. Unfortunately, two months after the Senate unanimously passed this bill, it failed to pass in the House of Representatives by one vote. Representative Graves, Chair of the House Transportation and Infrastructure Committee, a pilot, aircraft owner, and co-chair of the House General Aviation Caucus, opposes the ROTOR Act stating that it would be burdensome for General Aviation. In an interview, Representative Graves said, "[the ROTOR Act] is emotional legislation, and the worst legislation is legislation that comes about as a result of an accident like this. Nothing in this act would have prevented what happened to DC."

Representative Graves crafted a modified version of the Senate bill. His bill, known as the ALERT Act (Airspace Location and Enhanced Risk Transparency Act) would not require "certain" aircraft to be equipped with ADS-B In. Representative Graves says his bill will address all the NTSB recommendations, but NTSB Chair Jennifer Homendy said this" watered-down" bill won't do enough to prevent a tragedy.

I respectfully disagree with Representative Graves' position, and I am astounded that a politician, to whom aviation is a hobby, can blatantly disregard safety recommendations from the professionals who are tasked with making those recommendations. To say that nothing in the ROTOR Act would have prevented the DCA collision is absurd. Of course, there is no guarantee that ADS-B IN technology would have prevented this accident, just like there is no guarantee that wearing a seat belt will prevent a fatal injury, but the benefits of this system far outweigh the cost. Personally, having flown with ADS-B In and understanding its value, I would not fly without it any more than I would drive a car without wearing a seatbelt. Before Congress passes a bill that would exempt certain general aviation aircraft from complying with ADS-B In requirements, they should consider the following statistics. During a six-year period beginning in 2016, there were forty-three reports of midair collisions involving general aviation airplanes in the United States resulting in seventy-nine fatalities. Adding ADS-B In equipment to an airplane that is already equipped with ADS-B Out is relatively inexpensive, and the benefits will undoubtedly outweigh the cost.

While enhancements to aviation safety are debated, another fatal accident occurred. Just before midnight on March 22, 2026, an Air Canada CRJ collided with a fire truck at New York's LaGuardia airport fatally injuring both pilots and severely injuring a flight attendant. According to audio that has been released by multiple news outlets, the air traffic controller mistakenly cleared a fire truck to cross a runway just as the CRJ was touching down on that runway. Realizing his mistake, the controller tried in vain to stop the fire truck, but it was too late. Early reports indicate that a local controller and a controller-in-charge were the only occupants in the control tower at the time of the accident, which apparently is normal for the overnight shift. It is too early to determine if short staffing contributed to the accident, but minimal staffing was a contributing factor in the DCA collision.

The Laguardia control tower is equipped with Airport Surface Detection Equipment Radar which is used to identify traffic on the airport surface. The system, which has been in use for many years, tracks airplanes via their transponders. It can also detect non-transponder equipped vehicles, like fire trucks, but with reduced accuracy. Equipping ground vehicles with ADS-B Out transmitters would significantly improve accuracy and speed in detecting vehicles on the airport surface. In my opinion, the benefits of mandating ADS-B In for airplanes and ADS-B Out for ground vehicles far outweigh the cost. It has been reported that American Airlines has already equipped many of its airplanes with ADS-B In.

Moving to air traffic control, it would be easy to blame this accident on the lone local controller who failed to notify Captain Campos and First Officer Lilley about the approaching helicopter, as required by ATC rules. He made a deadly mistake, a mistake that will haunt him for the rest of his life, but you can't ignore the fact that he was put in an impossible position. Years earlier it was considered necessary to add a second local controller to handle the high volume of helicopter traffic near the airport. You cannot just abandon that position, increase the workload on the remaining controller, and hope that things work out. We are all human and humans make mistakes, but no mistake was greater than the decision to allow this position to be unmanned.

The lone controller was not negligent or careless, he was overloaded. He worked in a culture where he had to use every tool at his disposal to "make it work." Recall that the DCA Air Traffic Manager at the time of the accident testified that "it can be taxing on a person, you know, constantly having to give, give, give, or push, push, push...." DCA was a notoriously busy airport when I first flew there nearly 50 years ago and thanks to the relentless push by airlines to add slots, traffic congestion is worse than ever.

The automated Conflict Alert that was triggered in the control tower failed to prevent this accident. This system has no way of knowing that visual separation is being applied and it cannot be silenced so it becomes a "nuisance alert." That is, the alert continues to announce danger, but it is ignored because the threat has been eliminated by pilot provided visual separation. But, we know, from the tragic midair collision over San Diego in 1978, that a conflict alert requires positive action by the controller. How many times should we prove that? We have technology that is superior to the current radar-based system, but it will take years, and an Act of Congress, to implement this system. In the interim, proactive action must be taken by controllers whenever a conflict alert is triggered.

Airport congestion is central to this accident and if changes are not made it will contribute to the next. Controllers have repeatedly asked the FAA to reduce the rate of traffic arriving at DCA, but their concerns have been ignored. How can the agency that is tasked with regulating aviation, to promote safety, turn a blind eye to safety concerns?

DCA is inadequate for its current use. The cumulative effect of increasing the number of slots and expanding the perimeter rule beyond reasonable limits, the conflict created by military and civilian use of the airspace, poorly defined helicopter routes, airspace restrictions, short runways, a hazardous runway layout, antiquated air traffic technology, a shortage of qualified controllers, a revolving door of control tower managers, the use of improvised procedures to "make-it-work" and the refusal by the FAA to address controllers concerns, defines a culture that does not prioritize safety.

There are many reasons why this accident should not have happened, but one that stands out for me is "history repeating itself." Earlier, I described five fatal accidents that led to changes in the way we deice airplanes and there has not been a similar accident since that problem was properly addressed. The same holds true for thunderstorm related "microburst" accidents. Between 1975 and 1994, there were four fatal accidents attributed to a microburst during takeoff or landing. After studying the 1975 crash of Eastern 66 at New York's JFK airport, meteorologist Ted Fujita, the creator of the "F" scale which classifies the severity of tornados, proposed new methods to detect microbursts, but his ideas were not readily accepted. That is, until the1982 crash of Pan Am 759 in New Orleans followed by the 1985 crash of Delta 191 at Dallas/Fort Worth. This led to the development and widespread use of Doppler Radar, but not in time to prevent the 1994 microburst accident of US Air 1016 at my home base in Charlotte, North Carolina. Thanks to the development of Doppler Radar, there has not been a microburst accident since. This begs the question; how many accidents must we suffer before problems that plague DCA are addressed? Congress, and the stakeholders, can fix the problems that threaten DCA today, or they can cross their fingers and hope that history is not repeated.

Appendix A
Letter to Congress

A LETTER TO MEMBERS OF CONGRESS representing Virginia, Maryland, and the District of Columbia to leaders of the House Committee on Transportation and Infrastructure to express strong local opposition to "any changes to the current High Density ("Slot") and Perimeter rules at Ronald Reagan Washington National Airport.

Dear Chair Graves and Ranking Member Larsen:

As Congress begins consideration of a Federal Aviation Administration (FAA) reauthorization bill this year, we write to strongly oppose any changes to the current High Density ("Slot") and Perimeter rules at Ronald Reagan Washington National Airport ("National").

National and Washington Dulles International Airport ("Dulles") operate as an integrated system of federally owned assets. National was never intended to be a long-haul airport. The dual airport system was crafted with this is mind to accommodate limited land and runways at National. Dulles occupies 11,830 acres while National is just 860 acres. Acknowledging the physical limitations and community impacts of aircraft noise at National, Congress mandated the Slot and Perimeter rules.

National is currently designed to accommodate 15 million passengers annually. Last year, the airport set a record of 24 million passengers. Regarding safety, at its current level of activity, National already experiences an above average number of missed approaches and early turnouts because of weather, high demand, airfield layout, and runway length. Additional daily flights would likely increase the number of missed approaches and early turnouts, disrupting an already complicated airspace and impacting safety. Additionally, National has limited capacity of airport infrastructure such as gates, ticket counters, baggage handling areas, and parking, which are currently being stretched thin to handle the existing level of commercial flights. Adding slots or an expanding National's perimeter would further strain facilities at National.

Previous slot and perimeter changes have prevented Dulles from realizing its full potential as the primary long-haul flight destination for the Washington metropolitan area. The more expansive facilities at Dulles are structured to allow larger planes to land and take off, which yields efficiencies for customers, carriers, and our climate. In just the last decade, the federal government has spent significant amounts of money to bolster Dulles's infrastructure. For example, Phase 2 of the Washington Metropolitan Area Transit Authority's Metrorail Silver Line expansion opened less than six months ago and links the Washington metropolitan area to Dulles. The $6.8 billion, 41-mile-long Silver Line is Metro's largest expansion since its creation in 1976. Further, a recently announced $50 million grant from the Bipartisan Infrastructure Law will provide for construction of a new terminal at Dulles.

Our priority should be the safety and efficiency of flights, not the personal convenience of a comparatively small number of powerful and well-connected individuals. No Member of Congress appreciates another representative meddling with the assets in their state or district. We, too, strongly oppose any attempts by other Members and special interest groups to dictate operations at these airports for their own personal convenience at great cost to our communities and constituents.

For these reasons, we look forward to working with you to pass an FAA reauthorization bill this year that leaves intact the current rules governing operations at National.

Sincerely,

Rep. Don Beyer (VA-8)

Rep. Eleanor Holmes Norton (DC)

Rep. Jennifer Wexton (VA-10)

Rep. Gary Connolly (VA-11)

Rep. Steny Hoyer (MD-5)

Rep. Dutch Ruppersberger (MD-2)

Rep. David Trone (MD-6)

Rep. Abigail Spanberger (VA-7)

Rep. Jamie Raskin (MD-8)

Rep. John Sarbanes (MD-3)

Rep. Kweisi Mfume (MD-7)

Rep. Jennifer McClellan (VA-4)

Rep. Bobby Scott (VA-3)

Rep. Glen Ivey (MD-4)

Appendix B
NTSB Accident Summary

THIS REPORT DISCUSSES THE JANUARY 29, 2025, midair collision involving a Sikorsky UH-60L helicopter, operated by the US Army under the callsign PAT25, and a Mitsubishi Heavy Industries RJ Aviation (formerly Bombardier) CL-600-2C10 (CRJ700) airplane, N709PS, operated by PSA Airlines as flight 5342, over the Potomac River in southwest Washington, DC, about 0.5 miles southeast of Ronald Reagan Washington National Airport (DCA), Arlington, Virginia. The two pilots, two flight attendants, and 60 passengers on board the airplane and all three crewmembers on board the helicopter died. Both aircraft were destroyed as a result of the accident. Safety issues discussed in this report include:

- Helicopter route design surrounding DCA;
- The extensive use of pilot-applied visual separation and the inherent limitations of see-and-avoid, including when using night vision goggles;
- Unclear and inconsistent Federal Aviation Administration (FAA) guidance on helicopter route altitudes and boundaries and operators' misinterpretation of those altitudes;

- Limitations and gaps in the traffic awareness, alerting, and collision-avoidance technologies available to both aircraft;
- Risks associated with separate helicopter and airplane radio frequencies and blocked transmissions;
- Controller workload, position-combining, and communication practices;
- Deficiencies in FAA safety culture, facility-level oversight, and post-accident drug- and alcohol-testing procedures; and
- Shortcomings in FAA and US Army safety assurance and risk management processes, including lack of proactive data sharing and safety analysis to identify and mitigate midair collision risk in complex terminal environments. As a result of this investigation, the National Transportation Safety Board makes 33 recommendations to the FAA, 8 recommendations to the US Army, 5 recommendations to the Department of War Policy Board on Federal Aviation, 2 recommendations to the Department of Transportation (DOT), 1 recommendation to the DOT Office of the Inspector General, and 1 recommendation to the RTCA (Radio Technical Commission for Aeronautics).

Appendix C
NTSB Conclusions

1. The pilots of flight 5342 were certificated and qualified in accordance with federal regulations.

2. The pilots of flight 5342 were medically qualified for duty, and available evidence does not indicate that they were impaired by effects of medical conditions or substances at the time of the accident.

3. Review of the flight 5342 pilots' time since waking and sleep opportunities in the days before the accident indicated that the pilots were unlikely to have been experiencing fatigue.

4. The pilot, instructor pilot, and crew chief onboard PAT25 were qualified and current in their positions as designated by the unit commander in accordance with Army regulations.

5. The pilot, instructor pilot, and crew chief of PAT25 were medically qualified for duty, and available evidence does not indicate that they were impaired by effects of medical conditions or substances at the time of the accident.

6. Review of the three PAT25 crew members' time since waking and sleep opportunities in the days before the accident indicated that the crew were unlikely to have been experiencing fatigue.

7. The airplane was properly certificated, equipped, and maintained in accordance with 14 Code of Federal Regulations Part 121. The airplane was operated within its weight and balance limitations throughout the flight. Examination of the airplane revealed damage consistent with an in-flight collision and subsequent impact with water, and there was no evidence of any structural, system, or powerplant failures or anomalies. Review of surveillance videos indicated that the airplane's wing navigation, landing/taxi, and anti-collision strobe lights were operating at the time of the collision.

8. The helicopter was properly certificated, equipped, and maintained in accordance with US Army regulations. Review of helicopter maintenance records did not reveal any open discrepancies or anomalous trends that contributed to the accident. The helicopter was operated within its weight and balance limitations throughout the flight. Examination of the helicopter revealed damage consistent with an in-flight collision and subsequent impact with water, and there was no evidence of any structural, main or tail rotor system, flight control system, or powerplant failures or anomalies. Review of surveillance videos indicated that the helicopter's right and tail position lights, the landing light, as well as both upper and lower anti-collision lights, were operating at the time of the collision.

9. The operations supervisor and four controllers who were working in the Ronald Reagan Washington National Airport air traffic control tower cab at the time of the accident were properly certified, qualified in accordance with federal regulations and facility directives, and current.

10. Although the Ronald Reagan Washington National Airport air traffic control tower facility was not staffed to its target level at the time of the accident, the number of staff in the tower at the time of the accident was adequate and in accordance with Federal Aviation Administration directives.

11. The decision to combine the helicopter control and local control positions was not the result of insufficient staffing, and personnel were available to staff the helicopter control and local control positions separately had the operations supervisor chosen to do so.

12. The local control controller, assistant local controller, and operations supervisor were medically qualified for duty, and available evidence does not indicate they were impaired by effects of medical conditions at the time of the accident.

13. Review of the local control and assistant local control controllers' and operations supervisor's (OS) time since waking and sleep opportunities in the days before the accident indicated that the controllers, including the OS, were unlikely to have been experiencing fatigue.

14. Visual meteorological conditions prevailed in the area at the time of the accident. A review of observations recorded throughout the night of the accident revealed no evidence of any local atmospheric pressure anomalies that would have impacted barometric altimeter readings.

15. Metropolitan Washington Airports Authority aircraft rescue and firefighting and airport operations staff responded immediately and in accordance with applicable emergency plans and regulatory

requirements, deploying land- and water-based resources, and coordinating mutual aid under complex nighttime and on water conditions.

16. Keeping the helicopter control and local control positions continuously combined on the night of the accident increased the local control controller's workload and negatively impacted his performance and situation awareness.

17. Had the helicopter and local control positions been staffed separately, PAT25 might have received a more timely and effective traffic advisory.

18. The local control and helicopter control positions should have been separated at the time of the accident given traffic volume and complexity.

19. In the two minutes before the accident when traffic volume was increasing, the assistant local controller should have prioritized surveillance of aircraft in the air in order to assist the local controller, rather than diverting her attention to the lower priority task of documenting helicopter information, which could have been completed when traffic volume and complexity had subsided.

20. Due to extended time on position at the time of the collision and his complacency, the operations supervisor was likely experiencing reduced alertness and vigilance, which decreased his awareness of the operational environment and reduced his ability to proactively assess the risks posed by the traffic and environmental conditions at the time of the accident.

21. The lack of mandatory relief periods for supervisory air traffic control personnel is contrary to human factors research that shows clear performance deterioration in situations of prolonged time on task.

22. Although the local control controller provided an initial traffic advisory to the crew of PAT25 in accordance with Federal Aviation Administration Order Job Order 7110.65, he did not provide a corresponding advisory to the crew of flight 5342 regarding PAT25's location and intention, which could have increased situation awareness for the crew of flight 5342.

23. If the local control controller had issued a standard safety alert to the flight crews of either aircraft as prescribed in Federal Aviation Administration Order Job Order 7110.65, providing the conflicting aircraft's position and positive control instructions, the crew of either aircraft could have taken immediate action to avert the impending collision.

24. Initial and recurrent scenario-based training in threat and error management would help controllers identify and mitigate risks and strengthen situation awareness.

25. A risk assessment or decision making tool would likely have benefited the accident operations supervisor in identifying and mitigating the operational risk factors that were present on the night of the accident.

26. Due to degraded radio reception, the crew of PAT25 did not receive salient information regarding flight 5342's circling approach to runway 33.

27. The PAT25 instructor pilot did not positively identify flight 5342 at the time of the initial traffic advisory despite his statement that he had the traffic in sight and his request for visual separation.

28. With several other targets located directly in front of the helicopter represented by points of light with no other features by which to identify aircraft type, and without additional position information from the controller, the instructor pilot likely identified the wrong target.

29. Interference that obscured the controller's "circling to" call, the microphone keying that blocked the PAT25 crew from receiving the instruction to "pass behind," ambiguous visual cues, and the lack of an integrated traffic awareness and alerting system likely reinforced the PAT25 crew's expectation bias that the airplane was among the traffic approaching runway 1 and did not pose a conflict.

30. The absence of documented training on Ronald Reagan Washington National Airport's fixed-wing procedures and the mixed-traffic operating environment represented a safety vulnerability for Army flight crews operating in the Ronald Reagan Washington National Airport Class B airspace.

31. Due to additive allowable tolerances of the helicopter's pitot-static/altimeter system, it is likely that the crew of PAT25 observed a barometric altimeter altitude about 100 ft lower than the helicopter's true altitude, resulting in the crew erroneously believing that they were under the published maximum altitude for Route 4.

32. A recurrent task to verify the continued accuracy of recorded flight data for US Army aircraft would help ensure the data integrity needed to support quality assurance and safety programs and accident investigations.

33. The Federal Aviation Administration and the Army failed to identify the incompatibility between the helicopter routes' low

maximum altitudes and the error tolerances of barometric altimeters, which contributed to helicopters regularly flying higher than published maximum altitudes and potentially crossing into the runway 33 glidepath.

34. Pilots need all available information on the potential total error, allowed by design, that could occur in flight on an airworthy barometric altimeter.

35. The Army's post-installation functional check of the transponder on the accident helicopter was insufficient to detect that it was not broadcasting Automatic Dependent Surveillance–Broadcast Out.

36. The Army's lack of a recurrent transponder inspection procedure resulted in the incorrect aircraft address being transmitted by the accident helicopter's transponder, and the incorrect automatic dependent surveillance–broadcast settings on several other helicopters being undetected.

37. Because the APX-123A transponder is designed for use on multiple aircraft platforms, it is possible that incorrect settings may be present on other aircraft used throughout the Department of War armed services.

38. The crew of flight 5342 did not see the helicopter until it was too late to avoid a collision because of the high workload imposed during the final phase of their approach, and due to the helicopter's low conspicuity and lack of apparent motion.

39. Times of compacted demand as a result of air carrier scheduling practices increased operational complexity and required mitigations by controllers to maintain spacing and surface movement.

40. Ronald Reagan Washington National Airport air traffic control tower routinely received less than the requested miles in trail spacing from Potomac Consolidated Terminal Radar Approach Control, which increased controller workload by requiring them to generate additional spacing to prevent delays or gridlock.

41. The practice of "offloading" arrival traffic on approach to runway 1 by asking pilots if they could accept a circling approach to runway 33 was a routine mitigation strategy for Ronald Reagan Washington National Airport controllers to generate spacing that was not provided by Potomac Consolidated Terminal Radar Approach Control.

42. Time-based flow management, or metering, would provide Potomac Consolidated Terminal Radar Approach Control and Ronald Reagan Washington National Airport air traffic control tower with a consistent flow of traffic with more accurate spacing and greater predictability, thereby reducing controller workload.

43. Ronald Reagan Washington National Airport air traffic control tower has significant airspace, airfield, mixed fleet, and operations complexities that appear to be inconsistent with its current facility level classification.

44. The Federal Aviation Administration Air Traffic Organization failed to recognize external compliance verification results as indicators of systemic traffic management, volume, and flow issues at Ronald Reagan Washington National Airport for which controllers were required to compensate.

45. The longstanding practice of relying on pilot-applied visual separation (see-and-avoid) as the principal means of separating

helicopter and fixed-wing traffic in the Washington, DC, area by Ronald Reagan Washington National Airport air traffic control tower, the Army, and other helicopter operators led to a drift in operating practices among controllers and helicopter crews that increased the likelihood of a midair collision.

46. Reliance on pilot-applied visual separation (see-and-avoid) as a primary means of separating mixed traffic introduced unacceptable risk to the Ronald Reagan Washington National Airport Class B airspace.

47. Ronald Reagan Washington National Airport air traffic control tower's procedure of maintaining a discrete helicopter frequency when the local and helicopter control positions were combined decreased overall situation awareness for pilots operating in the area.

48. Providing controllers with additional salient cues regarding the perceived severity of a potential conflict would reduce controller cognitive load and would likely improve reaction time to the most critical conflict alerts.

49. There was no evidence that the local control controller, assistant local control controller, or operations supervisor were under the influence of alcohol or prohibited drugs at the time of the accident; however, evidence was substantially limited by the lack of postaccident alcohol testing, and evidence was of somewhat lower quality than it would have been if drug testing had been conducted sooner following the accident.

50. The Federal Aviation Administration Air Traffic Organization's (ATO) drug and alcohol testing determination did not meet Department of Transportation (DOT) timeliness requirements;

furthermore, the ATO's decision to not conduct drug testing as soon as possible after the testing determination, and to not conduct alcohol testing at all, violated DOT requirements.

51. The delayed and inappropriate drug and alcohol testing determination was due in part to the Air Traffic Organization's (ATO) determination process being inadequately designed to routinely meet Department of Transportation requirements for timely testing, and in part to ATO staff's incomplete understanding of those requirements.

52. Annual reviews of helicopter route charts as required by Federal Aviation Administration Order 7210.3DD would have provided an opportunity to identify the risk posed by the proximity of Route 4 to the runway 33 approach path, but there is no evidence to support that these reviews were being performed at Ronald Reagan Washington National Airport.

53. The information published by the Federal Aviation Administration regarding Washington, DC, area helicopter routes was insufficient to provide helicopter and fixed-wing operators with a complete understanding of the helicopter route structure and its lack of procedural separation from fixed-wing traffic.

54. Current aeronautical charting does not provide information on visual flight rules helicopter routes that may conflict or come in close proximity to approach and departure corridors, which reduces pilot situation awareness.

55. The lack of Automatic Dependent Surveillance–Broadcast (ADS-B) Out from the accident helicopter did not contribute to this accident, as the helicopter was still being tracked by radar, and ADS-B Out would not have provided improved traffic alerting for the Ronald

Reagan Washington National Airport controller or the crew of flight 5342 because the airplane was not equipped with ADS-B In.

56. The Army's standard operating procedures that prevent flight crews from enabling Automatic Dependent Surveillance–Broadcast (ADS-B) Out while in flight, when not performing sensitive missions that require ADS-B to be disabled, limit the visibility of military aircraft on collision avoidance technologies that leverage ADS-B information.

57. Although the airplane's traffic alert and collision avoidance system operated as designed, it was ineffective in preventing the collision because current activation criteria and resolution advisory inhibit altitudes.

58. Traffic advisory aural alerts that include additional information about the location of traffic could reduce the time pilots need to visually acquire target aircraft.

59. Had the airplane been equipped with an airborne collision avoidance system that used Automatic Dependent Surveillance–Broadcast In information to show directional traffic symbols, the crew of flight 5342 would have received enhanced information about the risk posed by the helicopter, which could have enabled them to take earlier action to avert the collision.

60. Although the pilot and instructor pilot onboard PAT25 were equipped with tablets that had the ability to display traffic transmitting Automatic Dependent Surveillance–Broadcast Out, it is unlikely that the pilots were using the tablets to monitor or identify traffic at the time of the accident due to the workload associated with low-altitude flight.

61. Technological advances since the development of traffic alert and collision avoidance system II operating standards may allow airborne collision avoidance system Xa with reduced inhibit altitudes to have an expanded alerting envelope while reducing nuisance alerts.

62. Although not yet commercially available, had the helicopter been equipped with airborne collision avoidance system Xr with integrated aural alerting, the crew could have received an alert regarding flight 5342 and could have taken action to avert the collision.

63. Multiple data sources provided evidence of midair collision risk between fixed-wing aircraft and helicopters at Ronald Reagan Washington National Airport, including on approach to runway 33, before this accident; however, the limited access to and use of available objective and subjective proximity data hindered industry and government stakeholders' ability to identify hazards and mitigate risk.

64. Improving stakeholder access to standardized and objective information about aircraft close proximity encounters for use in safety assurance processes would increase the likelihood of detecting and mitigating hazards before accidents occur.

65. The Federal Aviation Administration's lack of an established process to inform parties about their involvement in events such as near midair collisions or traffic alert and collision avoidance system resolution advisories reduces the likelihood of fully understanding and mitigating future midair collision risk.

66. The Federal Aviation Administration Air Traffic Organization was made aware of, and had multiple opportunities to identify, the risk

of a midair collision between airplanes and helicopters at Ronald Reagan Washington National Airport; however, their data analysis, safety assurance, and risk assessment processes failed to recognize and mitigate that risk.

67. The Federal Aviation Administration Air Traffic Organization's application of its safety management system did not effectively coordinate safety assurance and safety risk management activities with external stakeholders in the Ronald Reagan Washington National Airport Class B airspace.

68. Changes to Ronald Reagan Washington National Airport air traffic control tower's standard operating procedures prior to the accident removing the requirement for the operations supervisor (OS) to document the time and reason for combining or de-combining the helicopter control position in the facility log made it less likely that the OS would consider and evaluate the risks associated with combining or de-combining the position.

69. Safety risk management practices were not fully integrated into Ronald Reagan Washington National Airport air traffic control tower operations and did not identify or mitigate the operational challenges faced by controllers or the lack of guidance regarding operational risk assessments for controllers and supervisors.

70. Federal Aviation Administration Air Traffic Organization (ATO) management did not follow the tenets of safety management systems to support its workforce, encourage open communication, identify and mitigate risks, or foster a just culture, which eroded the overall safety culture within the ATO.

71. The Army did not have a flight safety data monitoring program for helicopters, and as a result, was unaware of routine altitude exceedances and related risks in the Ronald Reagan Washington National Airport terminal area.

72. The Army's safety reporting systems for pilots were not well utilized and did not provide the organization with information about close encounters between Army helicopters and other aircraft that were later found to have occurred frequently.

73. The Army's process for allocating resources to aviation safety management did not ensure the development of a robust safety management system for helicopter operations in the Washington, DC, area.

74. The Army's aviation safety system failed to consistently detect, interpret, and act on signals of latent hazards, resulting in degraded safety assurance, organizational learning, and safety culture.

Previously Issued Findings

75. Separation distances between helicopter traffic operating on Route 4 and aircraft landing on runway 33 as they existed at the time of the accident were insufficient and posed an intolerable risk to aviation safety by increasing the chances of a midair collision.

76. When Route 4 operations are prohibited as recommended in Safety Recommendation A-25-1, it is critical for public safety helicopter operations to have an alternate route for operating in and around Washington, DC, without increasing controller workload.

Appendix D
NTSB Probable Cause

THE NTSB DETERMINES THAT THE PROBABLE CAUSE of this accident was the Federal Aviation Administration's (FAA) placement of a helicopter route in close proximity to a runway approach path; their failure to regularly review and evaluate helicopter routes and available data, and their failure to act on recommendations to mitigate the risk of a midair collision near Ronald Reagan Washington National Airport (DCA); as well as the air traffic system's overreliance on visual separation in order to promote efficient traffic flow without consideration for the limitations of the see-and-avoid concept. Also causal was the lack of effective pilot-applied visual separation by the helicopter crew, which resulted in a midair collision. Additional causal factors were the tower team's loss of situation awareness and degraded performance due to the high workload of the combined helicopter and local control positions and the absence of a risk assessment process to identify and mitigate real-time operational risk factors, which resulted in misprioritization of duties, inadequate traffic advisories, and the lack of safety alerts to both flight crews. Also causal was the Army's failure to ensure pilots were aware of the effects of error tolerances on barometric altimeters in their helicopters, which resulted in the crew flying above the maximum published helicopter route altitude.

Contributing factors include:

- The limitations of the traffic awareness and collision alerting systems on both aircraft, which precluded effective alerting of the impending collision to the flight crews;
- An unsustainable airport arrival rate, increasing traffic volume with a changing fleet mix, and airline scheduling practices at DCA, which regularly strained the DCA air traffic control tower workforce and degraded safety over time;
- The Army's lack of a fully implemented safety management system, which should have identified and addressed hazards associated with altitude exceedances on the Washington, DC, helicopter routes;
- The FAA's failure across multiple organizations to implement previous NTSB recommendations, including Automatic Dependent Surveillance–Broadcast In, and to follow and fully integrate its established safety management system, which should have led to several organizational and operational changes based on previously identified risks that were known to management; and
- The absence of effective data sharing and analysis among the FAA, aircraft operators, and other relevant organizations.

Appendix E
NTSB Recommendations

AS A RESULT OF THIS INVESTIGATION, the National Transportation Safety Board makes the following new safety recommendations.

To the Federal Aviation Administration—

Develop and implement time-on-position limitations for supervisory air traffic control personnel, including guidance for district and facility level management to adapt these limitations to account for their own staffing and local standard operating procedures. (A-26-8)

Develop instructor-led, scenario-based training on threat and error management that trains controllers to continuously monitor their environment to more quickly and accurately identify threats; promote team communication to ensure that communications are clear, timely, and assertive; emphasize effective scanning habits; recognize patterns in the development of adverse events; and enhance decision-making under stress by developing habits that balance procedural compliance with problem solving to mitigate the risks of threats and errors, and provide this training to all air traffic controllers annually. (A-26-9)

Develop and implement a risk assessment tool for supervisors that incorporates the principles of threat and error management to assist in risk identification, mitigation, and operational decision making. (A-26-

10) Initiate rulemaking in 14 Code of Federal Regulations Part 93 Subpart K, High Density Traffic Airports, that prescribes air carrier operation limitations at Ronald Reagan Washington National Airport in 30-minute periods, similar to those imposed at LaGuardia Airport, to ensure that the airport does not exceed capacity and to mitigate inconsistent air carrier scheduling practices. (A-26-11)

Fully implement operational use of the time-based flow management system at Potomac Consolidated Terminal Radar Approach Control and its associated air traffic control towers. (A-26-12)

Reassess the Ronald Reagan Washington National Airport's airport arrival rate with special consideration to its airspace complexity, airfield limitations, mixed-fleet operations, and traffic volume. (A-26-13)

Require each Class B or Class C air traffic control tower facility to evaluate its existing miles-in-trail procedures or agreements to ensure that the spacing provided is appropriate for operational safety, and make the results publicly available. (A-26-14)

Define objective criteria for the determination of air traffic facility levels considering traffic and airspace volume, operational factors unique to each facility, and cost of living. (A-26-15)

Using the criteria established by Safety Recommendation A-26-15, determine whether the classification of the Ronald Reagan Washington National Airport's air traffic control tower as a level 9 facility appropriately reflects the complexity of its operations. (A-26-16)

Develop a new and comprehensive instructor-led, scenario-based training on the proper use of visual separation, both tower- and pilot applied. This training should include information on the inherent limitations of see and avoid, responsibilities when applying visual separation, and guidance for controllers on factors, such as current traffic volume, workload, weather or environmental factors, experience, and staffing, that should be considered when applying visual separation. Require this training for all controllers and include on a recurrent basis thereafter in annual simulator refresher training. (A-26-17)

Conduct a comprehensive evaluation, in conjunction with local operators, to determine the overall safety benefits and risks to requiring all aircraft to use the same frequency when the helicopter and local positions are combined in the Ronald Reagan Washington National Airport air traffic control tower. (A-26-18)

Implement anti-blocking technology that will alert controllers and/or flight crews to potentially blocked transmissions when simultaneous broadcasting occurs. (A-26-19)

Develop and implement improvements to the conflict alert system to provide more salient and meaningful alerts to controllers based on the severity of the conflict triggering the alert. (A-26-20)

Once the improvements to the conflict alert system discussed in Safety Recommendation A-26-20 are implemented, provide training to controllers on its use. (A-26-21)

Revise the Air Traffic Organization's initial event response procedures so that an appropriate on-site supervisor makes each post accident and post incident drug and alcohol testing determination, based on their assessment of whether the event meets testing criteria and which controllers had duties pertaining to the involved aircraft, without needing to wait for investigation or approval. (A-26-22)

At least annually, provide training on the revised post-accident and post incident drug and alcohol testing determination procedure discussed in Safety Recommendation A-26-22 to all staff who have responsibilities under that procedure; this training should include a post-learning knowledge assessment. (A-26-23)

Ensure that annual reviews of helicopter route charts are being conducted throughout the National Airspace System as required by Federal Aviation Administration Order. (A-26-24)

Conduct a safety risk management process to evaluate whether modifications to the remaining helicopter route structure in the vicinity of Ronald Reagan Washington National Airport are necessary to safely deconflict helicopter and fixed-wing traffic and provide the results to the National Transportation Safety Board. (A-26-25)

Amend your helicopter route design criteria and approval process to ensure that current and future route designs or design changes provide vertical separation from airport approach and departure paths. (A-26-26)

Once the criteria and approval process referenced in Safety Recommendation A-26-26, review all existing helicopter routes to ensure alignment with these updated criteria. (A-26-27)

Incorporate the lateral location and published altitudes of helicopter routes onto all instrument and visual approach and departure procedures to provide necessary situation awareness to fixed-wing operators of the risk of helicopter traffic operating in their vicinity. (A-26-28)

Modify airborne collision avoidance system traffic advisory aural alerts to include clock position, relative altitude, range, and vertical tendency. (A-26-29)

Require existing and new traffic alerting and collision avoidance system (TCAS) I, TCAS II, and airborne collision avoidance system X installations to integrate directional traffic symbols. (A-26-30)

Require all aircraft operating in airspace where Automatic Dependent Surveillance—Broadcast (ADS-B) Out is required to also be equipped with ADS-B In with a cockpit display of traffic information that is configured to provide alerting audible to the pilot and/or flight crew. (A-26-31)

Require the use of the appropriate variant of airborne collision avoidance system X on new production aircraft that are subject to traffic alert and collision avoidance system equipage regulations. (A-26-32)

Require existing aircraft that are subject to traffic alert and collision avoidance system equipage regulations be retrofitted with the appropriate variant of airborne collision avoidance system X. (A-26-33)

Evaluate the feasibility of decreasing the traffic advisory and resolution advisory inhibit altitudes in airborne collision avoidance system Xa to enable improved alerting throughout more of the flight envelope. (A-26-34)

If the evaluation resulting from Safety Recommendation A-26-34 finds that the inhibit altitudes can be safely decreased, require retrofitting of the applicable airborne collision avoidance system X variant incorporating the reduced traffic advisory and resolution advisory inhibit altitudes on all aircraft that are subject to traffic alert and collision avoidance system and equipage regulations. (A-26-35)

Require that all rotorcraft operating in Class B airspace be equipped with airborne collision avoidance system (ACAS) Xr technology once the ACAS Xr standard has been published. (A-26-36)

Create an objective definition of close proximity encounter and a public database of those encounters and their locations that can be used to monitor their prevalence and identify areas of potential traffic conflict for safety assurance and safety risk management. (A-26-37)

Develop and implement a process that will, in a timely manner, notify involved parties after events such as near midair collisions or traffic alert and collision avoidance system resolution advisory activations, such that notification occurs while relevant data remains available and before meaningful safety analysis, reporting, or corrective action is no longer practicable. (A-26-38)

Based on the results of the audit completed in accordance with Safety Recommendation A-26-56, ensure that all safety management system 303 Aviation Investigation Report AIR-26-02 functions and data sharing activities at all air traffic control facilities are conducted in collaboration with all relevant external stakeholders. (A-26-39)

Establish a requirement across all air traffic control tower standard operating procedures that the operations supervisor (OS) or controller in-charge (CIC) document in the daily facility log when any control position is combined with the local control position, or when the OS/CIC position is combined with a control position, along with a rationale for doing so. (A-26-40)

To the US Army—

Revise training procedures for flight crews assigned to operate in the Washington, DC, area to ensure that they receive initial and recurrent training on fixed-wing operations at Ronald Reagan Washington National Airport, including approach and departure paths, runway configurations, and the interaction of those traffic flows with published helicopter routes. (A-26-41)

Develop and implement a recurring procedure, at an interval not to exceed 18 months, to verify the continued accuracy of recorded flight data. (A-26-42)

Incorporate information within the appropriate operator's manual for all applicable aircraft on the potential total error allowed by design that could occur in flight on an otherwise airworthy barometric altimeter, including the increased position error associated with the external stores support system configuration. (A-26-43)

Develop and implement a transponder inspection procedure on all aircraft with transponders capable of transmitting Mode S and Automatic Dependent Surveillance—Broadcast (ADS-B) and operated in the National Airspace System (NAS), at least annually and upon each aircraft's entry into service in the NAS, that ensures 1) the transponder ADS-B settings are correct, 2) the transponder is transmitting ADS-B, and 3) the transponder is transmitting the correctly assigned address. (A-26-44)

Establish a flight data monitoring program for rotary-wing aircraft the US Army operates in the National Airspace System. (A-26-45) 304

Aviation Investigation Report AIR-26-02 Survey US Army helicopter pilots to identify barriers to the utilization of flight safety reporting systems, develop a plan to address the identified barriers, and implement that plan across Army aviation units. (A-26-46)

Revise the method for allocating resources to ensure the development of a robust safety management system that will, at a minimum, identify and monitor the potential for midair collisions between Army aircraft and civil air traffic operating in the National Airspace System. (A-26-47)

Develop and maintain a flight safety management capability that is independently resourced and functionally separate from its occupational and environmental health management system, and ensure that this capability is both culturally and functionally integrated with units conducting sustained flight operations in the National Airspace System. (A-26-48)

To the Department of War Policy Board on Federal Aviation—

Conduct a study to evaluate the quality of radio transmissions and reception for those aircraft operated within the National Airspace System to identify factors that degrade communications equipment performance and adversely affect the safety of civilian and military flight operations. (A-26-49)

Implement appropriate enhancements, based on the findings of the study recommended in Safety Recommendation A-26-49, to remediate identified deficiencies in air–ground radio communications performance. (A-26-50)

Require the Department of War to verify on all aircraft with transponders capable of transmitting Mode S and Automatic Dependent Surveillance—Broadcast (ADS-B) and operated in the National Airspace System (NAS), at least annually and upon each aircraft's entry into service in the NAS, that 1) the transponder ADS-B settings are correct, 2) the transponder is transmitting ADS-B, and 3) the transponder is transmitting the correctly assigned address. (A-26-51)

Require armed services to amend their operational procedures to allow flight crews to enable Automatic Dependent Surveillance—Broadcast Out while in flight. (A-26-52)

Require all military aircraft operating in the National Airspace System (NAS) be equipped with Automatic Dependent Surveillance—Broadcast (ADS-B) In with a cockpit display of traffic information that is configured 305 Aviation Investigation Report AIR-26-02 to provide alerting audible to the pilot and/or flight crew, and that such requirement apply wherever in the NAS the Federal Aviation Administration requires any aircraft to operate with ADS-B Out. (A-26-53)

To the Department of Transportation—

Require the Federal Aviation Administration to demonstrate at least annually that each air traffic control facility it operates has the routine capability to accomplish required post-accident and post incident drug and alcohol testing within the US Department of Transportation's specified timeframes of two hours for alcohol and four hours for drugs, and implement a process to ensure that any facility without such capability will demonstrate timely remediation. (A-26-54) Work with the Federal Aviation Administration (FAA) Administrator to convene an independent panel to conduct a comprehensive review of the safety culture within the FAA's Air Traffic Organization (ATO), and use the findings to enhance the

ATO's existing safety management system and integrate it into all levels of the organization. (A-26-55) To the Department of

Transportation Office of Inspector General—

Complete an audit of the Federal Aviation Administration (FAA) Air Traffic Organization's safety management system functions and data sharing activities at all air traffic control facilities and determine whether these activities are conducted in collaboration with all relevant external stakeholders, ensuring that the audit's results are documented, reported to the Secretary of Transportation and the FAA Administrator, and made available to the public. (A-26-56)

To the RTCA Program Management Committee—

Finalize and publish the minimum operational performance standards for airborne collision avoidance system Xr for rotorcraft. (A-26-57)

Previously Issued Recommendations On March 11, 2025, the NTSB issued a safety recommendation report titled Deconflict Airplane and Helicopter Traffic in the Vicinity of Ronald Reagan Washington National Airport, which issued the following urgent safety recommendations addressing the potential for midair collisions between helicopters operating on Route 4 and airplanes landing on runway 33 or departing from runway 15 at DCA identified during this investigation.

To the Federal Aviation Administration:

Prohibit operations on Helicopter Route 4 between Hains Point and the Wilson Bridge when runways 15 and 33 are being used for departures and arrivals, respectively, at Ronald Reagan Washington National Airport (DCA). (A-25-1) (Urgent)

Designate an alternative helicopter route that can be used to facilitate travel between Hains Point and the Wilson Bridge when that segment of Route 4 is closed. (A-25-2).

Appendix F
Acronyms

ACAS – Airborne Collision Avoidance System

ADS-B – Automatic Dependent Surveillance Broadcast

AFL-CIO – American Federation of Labor and Congress of Industrial Organizations

ALERT Act – Airspace Location and Enhanced Risk Transparency Act

ARTCC – Air Route Traffic Control Center

ATC - Air Traffic Control

ATO – Air Traffic Organization

BWI – Baltimore Washington International Airport

CAA – Civil Aeronautics Administration

CRJ – Canadair Regional Jet

DCA – Ronald Reagan Washington National Airport

EMAS – Engineered Materials Arresting System

FAA – Federal Aviation Administration

GPS – Global Positioning System

IAD – Washington Dulles International Airport

IFR – Instrument Flight Rules

ILS – Instrument Landing System

MWAA – Metropolitan Washington Airport Authority

NAS – National Airspace System

NASA – National Aeronautics and Space Administration

NATCA – National Air Traffic Controllers Association

NTSB – National Transportation Safety Board

NVG – Night Vision Goggles

OS – Operations Supervisor

PAPI – Precision Approach Path Indicator

PATCO – Professional Air Traffic Controllers Association

PSA – Pacific Southwest Airlines

RA – Resolution

ROTOR Act – Rotorcraft Operations Transparency and Oversight Reform Act

TA -Traffic Advisory

TCAS - Traffic Collision Avoidance System

TRACON - Terminal Radar Approach Control

VFR – Visual Flight Riles

VMC – Visual Meteorological Conditions

www.ingramcontent.com/pod-product-compliance
Lightning Source LLC
LaVergne TN
LVHW020510100826
845148LV00003B/746

9798234074591